EAI/Springer Innovations in Communication and Computing

Series Editor

Imrich Chlamtac, European Alliance for Innovation, Ghent, Belgium

The impact of information technologies is creating a new world yet not fully understood. The extent and speed of economic, life style and social changes already perceived in everyday life is hard to estimate without understanding the technological driving forces behind it. This series presents contributed volumes featuring the latest research and development in the various information engineering technologies that play a key role in this process. The range of topics, focusing primarily on communications and computing engineering include, but are not limited to, wireless networks; mobile communication; design and learning; gaming; interaction; e-health and pervasive healthcare; energy management; smart grids; internet of things; cognitive radio networks; computation; cloud computing; ubiquitous connectivity, and in mode general smart living, smart cities, Internet of Things and more. The series publishes a combination of expanded papers selected from hosted and sponsored European Alliance for Innovation (EAI) conferences that present cutting edge, global research as well as provide new perspectives on traditional related engineering fields. This content, complemented with open calls for contribution of book titles and individual chapters, together maintain Springer's and EAI's high standards of academic excellence. The audience for the books consists of researchers, industry professionals, advanced level students as well as practitioners in related fields of activity include information and communication specialists, security experts, economists, urban planners, doctors, and in general representatives in all those walks of life affected ad contributing to the information revolution.

Indexing: This series is indexed in Scopus, Ei Compendex, and zbMATH.

About EAI - EAI is a grassroots member organization initiated through cooperation between businesses, public, private and government organizations to address the global challenges of Europe's future competitiveness and link the European Research community with its counterparts around the globe. EAI reaches out to hundreds of thousands of individual subscribers on all continents and collaborates with an institutional member base including Fortune 500 companies, government organizations, and educational institutions, provide a free research and innovation platform. Through its open free membership model EAI promotes a new research and innovation culture based on collaboration, connectivity and recognition of excellence by community.

Sachin S. Kamble • Rahul S. Mor • Amine Belhadi
Editors

Digital Transformation and Industry 4.0 for Sustainable Supply Chain Performance

 Springer

Editors
Sachin S. Kamble (iD)
EDHEC Business School
Roubaix, France

Amine Belhadi (iD)
Rabat Business School, International
University of Rabat
Salé, Morocco

Rahul S. Mor
Department of Food Engineering
National Institute of Food Technology
Entrepreneurship and Management (NIFTEM)
Kundli
Sonipat, Haryana, India

ISSN 2522-8595 ISSN 2522-8609 (electronic)
EAI/Springer Innovations in Communication and Computing
ISBN 978-3-031-19713-0 ISBN 978-3-031-19711-6 (eBook)
https://doi.org/10.1007/978-3-031-19711-6

This Springer imprint is published by the registered company Springer Nature Switzerland AG
The registered company address is: Gewerbestrasse 11, 6330 Cham, Switzerland

Preface

The recent literature provides a lot of information on the various Industry 4.0 technologies such as the Internet of Things (IoT), additive manufacturing, big data analytics, cloud computing, and robotic systems. However, more insights are required by supply chain practitioners on how Industry 4.0 technologies drive digital transformation in manufacturing organizations and what impact it has on the overall supply chain performance. This book offers an interplay between digital transformation, Industry 4.0 technologies, and sustainable supply chain performance.

Chapter 1 reviews and classifies the literature on big data analytics (BDA) application in the supply chain (SC) and proposes a framework that identifies SC visibility as a key driving force for SC transformation, achieved through strong BDA capability. The applications of Industry 4.0 techniques in creating a more promising circular economy concept (CEC) in manufacturing industries to fulfill consumer demand and change consumer behavior for sustainability are investigated in Chap. 2. The challenges of adopting smart technologies, security threats, technology up-gradation, high investment, and partners' rejection of smart technologies interventions for sustainable agri-food supply chain are covered in Chap. 3. Chapter 4 proposes different linear formulations for the "Wireless Sensors Location Problems" in the context of Industry 4.0 and validated by solving small and medium data set instances using CPLEX Solver. The challenges of implementing smart, green, resilient, and lean practices are analyzed through the ISM-based modelling approach in Chap. 5. A blockchain-based framework suitable for the ASCM context is developed based on blockchain and machine learning to authenticate, validate, and secure transactions between suppliers and legitimate users in ASCM environments in Chap. 6. Chapter 7 discusses the role of IoT (Internet of Things) and IIoT (Industrial Internet of Things) in supplier and customer continuous improvement interface. Chapter 8 examines the role of artificial intelligence integrated with customer relationship management (AI-CRM) systems in organizations and how it can help their growth by improving sales performance and decision-making. Chapter 9 provides brief research and compares the conventional methods or models accessible in literature and the role of Industry 4.0. In Chap. 10, the authors optimize the available financial resources (government aid, internal, and external resources) with a data envelopment analysis method and incorporate ratio data with zero inputs. Chapter 11 discusses how artificial intelligence (AI) can effectively enhance food hygiene, safety and quality, efficient production, and supply chain by efficient decision-making, food waste management, and smart sorting and packaging solution through economic resource utilization by reducing errors and saving capital investments. In Chap. 12, a study has been conducted on the Industry 4.0–based agritech adoption in farmer producer organizations.

Roubaix, France   Sachin S. Kamble
Sonipat, Haryana, India   Rahul S. Mor
Salé, Morocco   Amine Belhadi

Acknowledgments

We acknowledge all those people who were involved and helped in completing this book project. Firstly, we thank the authors for contributing their valued time and expertise. Special thanks are due to the reviewers' valuable contributions regarding the improvement of quality, coherence, and content demonstration of the chapters. We also appreciate the referees for reviewing the manuscripts and scholars for editing and organizing the chapters. Finally, the editors are grateful to their parent institutes for providing essential support to complete this project.

Roubaix, France	Sachin S. Kamble
Sonipat, Haryana, India	Rahul S. Mor
Salé, Morocco	Amine Belhadi

Contents

Big Data Analytics for Supply Chain Transformation: A Systematic Literature Review Using SCOR Framework

Sachin S. Kamble, Rahul S. Mor, and Amine Belhadi

1 Introduction

The big data analytics (BDA) applications in SCM have received significant attention in the existing literature. Several reviews are published focusing on BDA on logistics and SCM listed in Table 1. Previous research has primarily focused on the classification of the BDA application areas based on the type of analytics used – descriptive, predictive and prescriptive (Nguyen et al., 2017; Barbosa et al., 2017; Wang et al., 2016c; Souza, 2014; Gandomi Haider, 2015) – operations and SCM functions (Lamba & Singh, 2017; Nguyen et al., 2017; Olson, 2015), logistics and SC strategy (Wang et al., 2016c), SCM resources and process (Barbosa et al., 2017) and BDA technologies (Zhong et al. 2016b). Souza (2014) categorized applications of SCA in terms of descriptive, predictive and prescriptive analytics along with the supply chain operations reference (SCOR) model domains plan, source, make, deliver and return. However, the focus of the study was more on how SCA may be applied across SCOR domains without explaining in detail the real-life applications in the context of SCOR domains. The researchers have not explored using the SCOR model domain classification scheme, viz. plan, procure, make, deliver and

S. S. Kamble (✉)
EDHEC Business School, Roubaix, France
e-mail: sachin.kamble@edhec.edu

R. S. Mor
Department of Food Engineering, National Institute of Food Technology
Entrepreneurship and Management, Sonepat, Haryana, India

A. Belhadi
Rabat Business School, International University of Rabat, Salé, Morocco

© The Author(s), under exclusive license to Springer Nature Switzerland AG 2023
S. S. Kamble et al. (eds.), *Digital Transformation and Industry 4.0 for Sustainable Supply Chain Performance*, EAI/Springer Innovations in Communication and Computing, https://doi.org/10.1007/978-3-031-19711-6_1

Table 1 Literature reviews on BDA in SCM

Authors	#Articles	Time range	Analysis/categorization	Research objective
Arunachalam et al. (2017)	82	2008–2016	Bibliometric and thematic analysis	Development of BDA capabilities maturity model
Lamba and Singh (2017)	–	–	O&SCM functions	Identify future perspectives
Nguyen et al. (2017)	88	2011–2017	i. BDA application areas of SCM ii. Level of analytics (descriptive, predictive and prescriptive) iii. BDA models iv. BDA techniques	Future research directions
Barbosa et al. (2017)	44	2005–2016	i. Level of analytics ii. Use of SCM resources iii. SCM processes	To identify how significant BDA is to support value achievement
Wang et al. (2016c)	104	2004–2014	Level of analytics (descriptive, predictive, prescriptive)	To develop a supply chain analytics maturity framework of SCA based on functional, process-based, collaborative, agile SCA and sustainable SCA capability levels
Addo-Tenkorang and Helo (2016)		2010–2015	General classification based on BD technologies and BDA in SCM.	Identifying issues and proposal of a value-adding framework
Wamba et al. (2015)	62	2006–2012	Dimensions related to business value creation from BD	Assessment of the business value of BD
Olson (2015)	–	–	SCM functions	To observe recent trends and developments, problems and opportunities
Gandomi and Haider (2015)	–	–	Types of BD analytics	To describe, review and reflect on BDA
Mishra et al. (2016)	57	2011–2015	Bibliometric analysis of BDA applications in SCM	To identify current trends and future directions for research
Souza (2014)	Nil	Nil	Level of BDA across SCOR model domains	Possible applications of advanced analytics
Zhong et al. (2016a)	–	–	Representative BDA applications in SCM	To review the current movement of BDA applications in SCM and identify current challenges, opportunities and future perspectives

return. In the literature, the SCOR model is a very well-recognized SC model used by practitioners in different industries worldwide (Lockamy & McCormack, 2004; Theeranuphattana & Tang, 2008; Souza, 2014).

Classification of BDA applications across the SCOR domains will benefit SCM practitioners (Huang et al., 2005). BDA covers the comprehensive capability for the interface between IT assets and other firm resources and is considered an organizational capability (Cosic et al., 2015; Barbosa et al., 2017). SCM practitioners may be interested in understanding which resources BDA uses across the SCOR domains. A literature review focusing on BDA's role in the better use of SCM resources is scarce (Barbosa et al., 2017). So, this chapter uses a classification framework, which categorizes and connects the SCOR domains with the analytics and SCM resources level. The amount of investment to be made on BDA initiatives is difficult for an organization. An organization may not be able to invest equally in all the areas of the supply chain. Information about the latest BDA applications in SCM across the SCOR domains and the significant supply chain resources used for data collection, analysis and data-driven decision-making will be beneficial for the organizations to make informed decisions, prioritizing the areas of BDA investment. Therefore, this review aims to provide useful insights on how BDA is applied in SCM by mapping BDA techniques and SCM resources to SCOR domains, viz. plan, source, make, deliver and return.

Section 2 of this chapter describes the research methodology and process used in the study (i.e. systematic mapping). The results are presented in Sect. 3. In Sect. 4, an SC visibility and BDA capability framework is proposed based on the findings of the categorical structure used in the study. The areas identified for future research are presented in Sect. 5. Section 6 offers the conclusions and limitations of the research.

2 Review Methodology

The review methodology adopted in this chapter is based on the content analysis approach proposed by Mayring (2003). A similar review approach was used by Seuring and Muller (2008), Govindan et al. (2014), Gao et al. (2016) and Arunachalam et al. (2017). The review was systematically conducted using a four-step iterative process. The steps included: material collection, descriptive analysis, category selection and materials evaluation.

2.1 Material Collection

To have an all-inclusive coverage of all the possible applications of BDA in logistics and SCM, the following keywords were used: BD, data mining, analytics, business intelligence, data-driven, predictive analytics, real-time data, forecasting, product development, sourcing, procurement, production, logistics, SCM, inventory, maintenance, quality, operations, innovation, order picking, transportation and manufacturing. The timeline starting from 2010 was selected for review as BDA has become a global phenomenon only after 2010, with significant research in this field regarding the volume and considerable contributions to theory and practices (Manyika et al., 2011; Nguyen et al., 2017). Further, this chapter reviews the significant developments in BDA applications in SCM through an 8-year analysis. The initial search generated 1657 papers, and 1021 were retained after removing the duplication and verifying in the Mendeley software. The inclusion criteria ensured that research from academic sources, such as peer-reviewed journals and reputed conference proceedings, was selected. The review did not include any publications in books, book chapters, doctoral work, white papers, editorial columns, etc. With the application of the inclusion criteria, the total number of papers dropped to 634. Upon meeting the inclusion criteria, the selected articles were subjected to exclusion criteria to reduce the number of papers by removing the publications that were not aligned with the scope of the present literature review. The exclusion criteria applied included reading the introduction and conclusion of each paper and removing those publications that do not deal with the application or benefits of the BDA in SCM (Lamba & Singh, 2017; Nguyen et al., 2017; Arunachalam et al., 2017). It was observed that many articles had the term BDA mentioned in the body of text or just pointed out potential benefits in the field of SCM without detailing how BDA is applied or how it can be implemented. Such papers were excluded from this study bringing the final count of publications to 220 for a full review.

2.2 Descriptive Analysis

Figure 1 shows that the papers published on BDA applications in SCM have continuously increased over the last 7 years. The trend has seen an enormous increase since 2014. The publication frequency distribution aligns with Gandomi

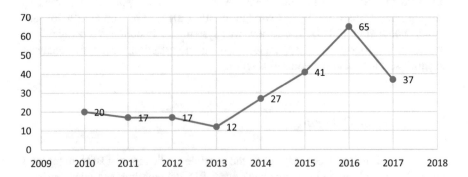

Fig. 1 Articles published from 2010 to 2017

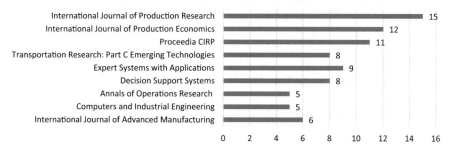

Fig. 2 Distribution of journals publishing the referred articles

and Haider (2015) and Nguyen et al. (2017). It indicates that the BDA application in SCM is gaining much attention from researchers. The selected 220 papers are from 87 different journals; only nine published more than five papers (Fig. 2).

2.3 Article Classification

The classification step conceptualizes the framework through structural dimensions and analytics, i.e. SCOR domains, level of analytics and SCM resources. The SCOR model developed by Supply Chain Council has five main processes: plan, source, make, deliver and return.

(a) Plan: This process analyses the information and forecasts the market trends of goods and services.
(b) Source: This process deals with the procurement system. It includes the activities related to ordering and receiving materials and products. Major decisions include supplier selection, negotiations, vendor management and evaluations.
(c) Make: This process in the model covers the manufacturing of goods. The related activities are scheduling, manufacturing, repairing, remanufacturing and recycling materials and products.
(d) Deliver: In this process, the movement of the finished goods and services to reach planned or actual demand is covered. The related activities are receiving, scheduling, picking, packing and shipping orders.
(e) Return: It is processed, returning the goods or receiving the product in a reversed loop. The related activities are to request, approve and determine the disposal of products and assets (Lockamy & McCormack, 2004; Trkman et al., 2010; Delipinar & Kocaoglu, 2016; Souza, 2014).

The categories used for the level of analytics were descriptive, predictive and prescriptive analytics. This classification is notable in literature (Wang et al., 2016c; Barbosa et al., 2017; Nguyen et al., 2017). The categorization suggested by

Braganza et al. (2017) was used for the third layer, namely, SCM resources. This classification scheme was used by Barbosa et al. (2017) to review the extent of using BDA to manage SCM resources. The descriptive statistics based on the classification scheme are presented below.

2.3.1 Classification by SCOR Domains

The distribution of the selected papers for the review across the SCOR domains, viz. plan, source, make, deliver and return, is presented in Fig. 3. If any studies did not relate to one or more specific domains, they were classified under 'Overall Supply Chain' (OSC). As seen from Fig. 3, plan, make and deliver domains have dominated the current literature, taking up to 76% of the publications.

Research on the source domain is limited to 10% publications. However, the return domain has not received much attention from the researchers, with no publications from the selected list of articles. Figure 4 presents the trend of research papers on how BDA is applied across the SCOR domains between 2010 and 2017. It is observed that there were very few papers before 2013. The real trend starts from 2013 onwards. There is an increased interest in the plan, make and deliver domain from 2013 onwards, with high growth in the papers of making domain from 2015 surpassing other domains. The research in the source domain continues to grow at a steady rate.

Fig. 3 Classification of articles by SCOR domain

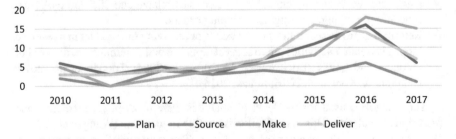

Fig. 4 Trend of publications on SCOR domain

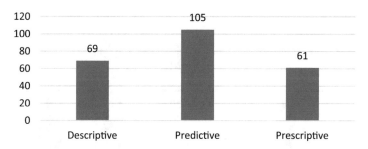

Fig. 5 Classification of papers on level of analytics

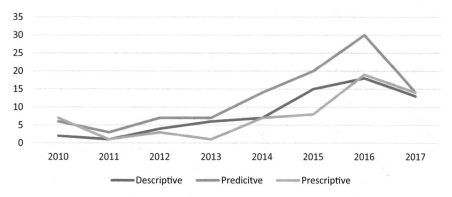

Fig. 6 Trend of publications on level of analytics

2.3.2 Classification by Level of Analytics

Figure 5 depicts the distribution of use of different levels of BDA in the literature. It is observed that predictive analytics is used in most of the studies (47%), followed by descriptive analytics (31%) and prescriptive analytics (27%). Figure 6 illustrates the popularity of each analytics type by year. The use of predictive analytics is seen to be dominating from 2010 to 2017, with all three types of analytics having an increasing trend.

2.3.3 Classification by SCM Resources

Figure 7 distributes the papers for SCM resources managed by BDA. Most papers address organisational and technological resources (56 and 54%, respectively) because BDA deals with a significant amount of data. Data, information and knowledge considered organizational resources provide rich and valuable insights into decision-making.

The data are collected and stored from different sources in all the domains, and meaningful insights are extracted. Further, this voluminous data requires advanced storage and retrieval systems like ERP, RFID, Cloud servers, etc. Therefore,

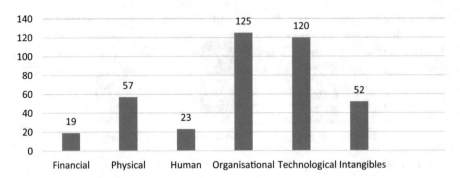

Fig. 7 Categorization of articles on SCM resources

technological resources have comprehensive coverage in the BDA literature. The organizational resources with efficient data collection and storage technology are required to use BDA efficiently. Hence, we find that the technological resources are well spread across all the domains.

The physical resources have been studied in 25% of the studies. These studies include the processes that use physical resources such as physical manufacturing systems, movement of goods, order picking, storage and transportation. Intangible resources are covered in 23% of the studies. These resources are primarily associated with the new product development and innovation process. These studies capture the BD from the customers through product feedback, complaints, preferences, etc., providing them with an opportunity to act as value co-creators. Human and financial resources are the least covered area, which is addressed in 10% and 8% of the studies. With the increase in the use of BDA for solving SCM issues, there are concerns about the availability of technical skills to manage and analyse the data (Richey et al., 2016; Schoenherr & Speier-Pero, 2015). Human resources also include the studies where the employee responses in the form of surveys are captured for decision-making. Achieving financial performance is an objective of some of the studies. BDA applications in this area are not seen much. A few studies discuss optimizing the processes to reduce cost and increase profitability.

3 Results and Discussions

This section discusses the status of extant research about BDA applications in SC and logistics management. The results of the SLR are shown in Tables 2, 3 and 4. Table 2 presents the categorization scheme applied to all the selected papers. Table 3 presents the summarized details of the level of analytics across the SCOR domains, and Table 4 presents the resources managed by BDA across the SCOR domains. The SLR presented in Table 2 classifies the selected 220 papers into three main categories, i.e. level of analytics, supply chain resources used for BDA in SCM (viz.

Table 2 Systematic literature review

Sr. no.	Authors	Level of analytics			SCM resources used						SCOR domains				
		Descriptive	Predictive	Prescriptive	Financial	Physical	Human	Organizational	Technological	Intangible	Plan	Source	Make	Deliver	OSC
1	Addo-Tenkorang anf Helo (2016)	✓						✓	✓						✓
2	Ahiaga-Dagbui and Smith (2014)			✓	✓			✓					✓		
3	Akter et al. (2016)		✓						✓	✓		✓			✓
4	Alkhalifah and Ansari (2016)			✓					✓		✓				
5	Aloysius et al. (2016)			✓				✓	✓		✓				
6	Amarouche et al. (2015)	✓							✓	✓	✓				
7	Amos et al. (2016)			✓		✓							✓		
8	Arias and Bae (2016)		✓			✓		✓	✓		✓				
9	Arief et al. (2016)		✓					✓				✓			
10	Arunachalam et al. (2017)	✓	✓	✓				✓	✓						✓
11	Azadnia et al. (2013)			✓		✓		✓						✓	
12	Babiceanu and Seker (2016)		✓			✓		✓	✓				✓		
13	Bag (2016)			✓			✓		✓			✓	✓		
14	Bahrami et al. (2012)	✓						✓	✓		✓				
15	Balaban et al. (2015)	✓				✓			✓					✓	
16	Barbosa et al. (2017)	✓	✓	✓	✓	✓	✓	✓	✓			✓			✓
17	Bauer et al. (2016)			✓				✓					✓		
18	Bendoly (2016)	✓						✓			✓				
19	Bendoly et al. (2012)			✓						✓	✓				

(continued)

Table 2 (continued)

Sr. no.	Authors	Level of analytics			SCM resources used						SCOR domains				
		Descriptive	Predictive	Prescriptive	Financial	Physical	Human	Organizational	Technological	Intangible	Plan	Source	Make	Deliver	OSC
20	Berengueres and Efimov (2014)		✓					✓		✓	✓				
21	Bhattacharjya et al. (2016)	✓								✓	✓				
22	Blackburn et al. (2017)	✓						✓	✓		✓				
23	Bradley et al. (2017)	✓	✓		✓				✓		✓				
24	Brandenburger et al. (2016)		✓						✓				✓		
25	Brinch et al. (2017)	✓						✓	✓						✓
26	Butler and Bright (2014)			✓									✓		
27	Cárdenas-Benítez et al. (2016)		✓			✓		✓						✓	
28	Chae (2015)		✓			✓				✓	✓				
29	Chae and Olson (2013)		✓						✓						✓
30	Chae et al. (2014)		✓					✓							✓
31	Charaniya et al. (2010)		✓					✓					✓		
32	Chen and Blue (2010)		✓					✓			✓				
33	Chen et al. (2010)			✓	✓			✓	✓		✓				
34	Chen et al. (2016)		✓					✓	✓				✓		
35	Cheng et al. (2017)	✓						✓					✓		
36	Chiang et al. (2011)					✓		✓						✓	
37	Chien et al. (2013)		✓										✓		
38	Chien et al. (2017)		✓					✓					✓		

		C1	C2	C3	C4	C5	C6	C7	C8	C9	C10	C11	C12	C13	C14	C15	C16	C17
39	Choi et al. (2016)	✓			✓				✓					✓				
40	Chong et al. (2016)		✓						✓		✓		✓	✓			✓	
41	Chuang et al. (2014)		✓		✓													✓
42	Chung and Tseng (2012)		✓					✓	✓		✓		✓					
43	Chung and Tseng (2012)	✓						✓			✓		✓	✓				
44	Çiflikli and Özyirmidokuz (2010)			✓	✓				✓				✓		✓			
45	Cochran et al. (2016)		✓	✓					✓						✓			
46	Cohen et al. (2017)			✓					✓						✓			
47	Colace et al. (2014)	✓					✓	✓			✓		✓					
48	Cristobal et al. (2015)		✓				✓		✓						✓		✓	
49	Cui et al. (2016)		✓				✓		✓		✓						✓	
50	Davis et al. (2012)		✓						✓								✓	
51	Delen and Demirkan (2013)	✓						✓	✓									✓
52	Delen et al. (2011)		✓				✓	✓	✓		✓		✓				✓	
53	Diana (2012)		✓				✓	✓	✓								✓	
54	Dietrich et al. (2012)		✓		✓			✓	✓		✓		✓					
55	Djatna and Munichputranto (2015)		✓			✓			✓		✓				✓			
56	Dobre and Xhafa (2014)		✓					✓	✓		✓							✓
57	Dubey et al. (2016)	✓					✓		✓		✓		✓					
58	Dudas et al. (2014)			✓					✓			✓	✓					
59	Durán et al. (2010)			✓	✓				✓		✓	✓	✓					

(continued)

Table 2 (continued)

Sr. no.	Authors	Level of analytics			SCM resources used						SCOR domains				
		Descriptive	Predictive	Prescriptive	Financial	Physical	Human	Organizational	Technological	Intangible	Plan	Source	Make	Deliver	OSC
60	Ehmke et al. (2016)			✓										✓	
61	Eidizadeh et al. (2017)	✓						✓			✓				
62	Fiosina et al. (2013)	✓						✓	✓					✓	
63	Gandomi and Haider (2015)	✓	✓					✓	✓						✓
64	Gerunov (2016)	✓							✓	✓	✓				
65	Groves et al. (2014)			✓				✓				✓			
66	Guo et al. (2014)			✓				✓						✓	
67	Haberleitner et al. (2010)		✓		✓			✓	✓		✓				
68	Hammer et al. (2017)			✓					✓		✓		✓		
69	Haverila and Ashill (2011)	✓						✓		✓	✓				
70	Hazen et al. (2014)		✓					✓							✓
71	Hazen et al. (2016)	✓	✓	✓				✓	✓		✓				✓
72	He et al. (2015)		✓					✓	✓	✓	✓	✓			
73	Ho and Shih (2014)		✓						✓			✓			
74	Hofmann (2017)	✓						✓	✓	✓				✓	
75	Hsu et al. (2015)		✓							✓				✓	
76	Huang and van Mieghem (2014)			✓						✓				✓	
77	Ilie-Zudor et al. (2015)		✓			✓		✓			✓			✓	
78	Ivanov (2017)			✓				✓	✓			✓			
79	Jain et al. (2014)	✓						✓				✓			
80	Jain et al. (2017)			✓					✓				✓		

#	Reference																
81	Jeeva and Dickie (2012)	✓						✓		✓							✓
82	Jelena and Fiosins (2017)		✓				✓		✓			✓	✓		✓		
83	Jeon and Hong (2016)				✓		✓		✓		✓			✓	✓		
84	Ji-fan Ren et al. (2017)			✓	✓				✓		✓			✓			✓
85	Jin et al. (2016)		✓		✓						✓		✓		✓	✓	
86	Jun et al. (2014)			✓							✓					✓	
87	Kache and Seuring (2017)	✓	✓		✓				✓		✓			✓			✓
88	Kargari and Sepehri (2012)	✓	✓						✓		✓		✓				✓
89	Kemp et al. (2016)		✓						✓		✓		✓		✓		✓
90	Kibira (2015)		✓	✓					✓		✓	✓		✓	✓		
91	Kok and Shang (2014)				✓		✓		✓		✓		✓				✓
92	Köksal et al. (2011)	✓					✓			✓		✓		✓	✓		
93	Koo et al. (2015)	✓			✓			✓		✓					✓	✓	
94	Kowalczyk and Buxmann (2015)		✓		✓				✓		✓		✓	✓			
95	Kretschmer et al. (2017)		✓		✓		✓		✓		✓			✓			✓
96	Krumeich et al. (2016)		✓			✓					✓			✓	✓		✓
97	Kubac (2016)		✓			✓			✓		✓			✓	✓		
98	Kubáč (2016)		✓			✓	✓		✓					✓		✓	
99	Kuester and Rauch (2016)	✓								✓	✓		✓	✓			
100	Kumar et al. (2016)		✓						✓		✓			✓	✓		
101	Kumar et al. (2017)			✓			✓		✓		✓			✓	✓		

(continued)

Table 2 (continued)

Sr. no.	Authors	Level of analytics			SCM resources used						SCOR domains				
		Descriptive	Predictive	Prescriptive	Financial	Physical	Human	Organizational	Technological	Intangible	Plan	Source	Make	Deliver	OSC
102	Kuo et al. (2015)			√	√			√				√			
103	Kwak and Kim (2012)			√					√				√		
104	Lade et al. (2017)		√				√		√				√		
105	Lanka and Jena (2014)		√						√					√	
106	Lau et al. (2014)		√						√	√	√				
107	Lee (2016)			√				√						√	
108	Lee and Chang (2010)	√					√	√		√	√				
109	Lee et al. (2013)		√					√	√				√		
110	Lee et al. (2017)			√		√		√	√					√	
111	Levner et al. (2011)		√					√	√					√	
112	Li et al. (2015)		√					√							
113	Li et al. (2016a)		√					√	√	√	√				
114	Li et al. (2016b)		√			√								√	
115	Li et al. (2016c)		√					√	√				√		
116	Li et al. (2014)		√			√		√	√				√		
117	Lin et al. (2010)			√	√			√		√					
118	Liu et al. (2016)	√						√	√		√				
119	Ma et al. (2014)		√					√	√		√		√		
120	Mariadoss et al. (2014)		√				√	√	√		√				
121	Marine-Roig and Clavé (2015)		√						√	√	√				
122	Markham et al. (2015)	√							√	√	√				
123	Mason et al. (2017)		√					√	√				√		
124	Mehmood (2017)			√					√					√	
125	Min (2010)			√				√			√	√		√	

		C1	C2	C3	C4	C5	C6	C7	C8	C9	C10	C11	C12	C13	C14
126	Miroslav et al. (2014)		✓					✓	✓	✓			✓		
127	Mishra et al. (2016)	✓	✓	✓					✓	✓					✓
128	Miyaji (2015)		✓				✓	✓						✓	
129	Mori et al. (2012)	✓	✓					✓	✓	✓			✓		
130	Mourtzis et al. (2016)	✓		✓			✓		✓	✓		✓			
131	Moyne et al. (2016)		✓						✓	✓		✓			
132	Munro and Madan (2016)				✓					✓		✓			
133	Nguyen et al. (2017)	✓	✓		✓			✓	✓	✓					✓
134	O'Brien et al. (2014)	✓					✓	✓						✓	
135	Olson (2015)	✓	✓		✓				✓	✓					✓
136	Oruezabala and Rico (2012)	✓	✓						✓	✓	✓		✓		
137	Ostrowski et al. (2016)	✓						✓	✓	✓					✓
138	Packianather et al. (2017)	✓	✓			✓		✓	✓	✓					
139	Pang and Chan (2017)		✓				✓	✓	✓	✓				✓	
140	Papadopoulos et al. (2017)	✓						✓	✓				✓		✓
141	Park et al. (2016)	✓	✓					✓	✓	✓			✓	✓	
142	Peters and Link (2010)	✓							✓	✓				✓	
143	Petri et al. (2016)	✓	✓				✓		✓	✓	✓			✓	
144	Prasad et al. (2016)			✓		✓			✓	✓					
145	Ralha et al. (2012)		✓		✓			✓	✓	✓					
146	ur Rehman et al. (2016)	✓						✓	✓	✓			✓		
147	Reuter et al. (2016)	✓							✓	✓		✓			
148	Richey et al. (2016)	✓					✓		✓				✓		✓

(continued)

Table 2 (continued)

Sr. no.	Authors	Level of analytics			SCM resources used						SCOR domains				
		Descriptive	Predictive	Prescriptive	Financial	Physical	Human	Organizational	Technological	Intangible	Plan	Source	Make	Deliver	OSC
149	Robinson et al. (2015)			√		√							√		
150	Ronowicz et al. (2015)			√				√					√		
151	Salehan and Kim (2016)		√					√		√	√				
152	Sanders (2016)	√						√	√						√
153	Sangari and Razmi (2015)		√				√	√	√	√					√
154	Sann (2013)	√						√	√	√	√				
155	Schmidt et al. (2017)		√			√			√				√		
156	Schoenherr and Speier-Pero, (2015)	√					√	√	√	√					
157	Schoenherr and Swink (2015)		√						√	√	√				
158	Shafiq et al. (2017)	√						√	√				√		
159	Shan and Zhu (2015)		√			√			√					√	
160	Shanmugasundaram and Paramasivan (2016)		√						√	√	√				
161	Shi and Abdel-Aty (2015)		√			√			√					√	
162	Shin et al. (2014)		√		√			√			√		√		
163	Shukla and Kiridena (2016)			√				√			√			√	
164	Sivamani et al. (2014)		√			√	√		√					√	
165	Soban et al. (2016)			√				√					√		
166	Sodhi and Tang (2011)			√		√					√				

#	Reference																
167	Soroka (2017)									✓			✓				✓
168	Souza (2014)	✓						✓		✓			✓					
169	van der Spoel et al. (2017)	✓			✓								✓				✓	
170	Srinivasan and Swink (2017)	✓		✓	✓		✓		✓			✓	✓					✓
171	St. Aubin et al. (2015)	✓			✓		✓					✓	✓	✓			✓	
172	Stefanovic (2015)	✓			✓			✓						✓			✓	
173	Tachizawa et al. (2015)			✓	✓		✓											✓
174	Tan et al. (2015)	✓			✓		✓					✓	✓	✓				
175	Tanev et al. (2015)	✓			✓		✓		✓			✓	✓	✓				
176	Thiruverahan and Subramanian (2015)	✓							✓								✓	
177	Thotappa and Ravindranath (2010)			✓				✓									✓	
178	Toole et al. (2015)	✓							✓								✓	
179	Trkman et al. (2010)	✓							✓									✓
180	Tsai and Huang (2015)	✓			✓		✓		✓			✓					✓	
181	Tsao (2017)			✓			✓		✓			✓		✓				
182	Tsuda et al. (2015)	✓			✓				✓					✓				
183	Tu et al. (2016)	✓		✓	✓				✓			✓					✓	
184	Ulrike et al. (2013)	✓						✓	✓			✓	✓				✓	
185	Unay and Zehir (2012)	✓				✓		✓	✓				✓				✓	
186	Veugelers et al. (2010)	✓						✓	✓			✓	✓				✓	
187	Walker and Brammer (2012)			✓										✓				
188	Walker and Strathie (2016)	✓			✓				✓								✓	

(continued)

Table 2 (continued)

Sr. no.	Authors	Level of analytics			SCM resources used						SCOR domains				
		Descriptive	Predictive	Prescriptive	Financial	Physical	Human	Organizational	Technological	Intangible	Plan	Source	Make	Deliver	OSC
189	Wallander and Makitalo (2012)		✓			✓		✓	✓					✓	
190	Waller and Fawcett (2013)		✓				✓					✓			✓	
191	Wamba et al. (2015)	✓							✓						✓
192	Wang and Yang (2016)	✓				✓		✓	✓					✓	
193	Wang and Zhang (2016)		✓					✓			✓		✓		
194	Wang et al. (2017)		✓						✓				✓		
195	Wang et al. (2016a)		✓			✓		✓					✓		
196	Wang et al. (2016b)		✓						✓					✓	
197	Wang et al. (2016c)	✓	✓	✓		✓		✓	✓						✓
198	Wang et al. (2016d)			✓				✓	✓				✓		
199	Westerski et al. (2015)		✓					✓				✓			
200	Wiener and Julia (2010)		✓					✓					✓		
201	Wu et al. (2013)	✓				✓		✓				✓		✓	
202	Wu et al. (2017a, b)		✓	✓				✓	✓						✓
203	Xiao et al. (2015)	✓							✓	✓	✓			✓	
204	Xie (2016)		✓						✓					✓	
205	Xu and Guting (2013)		✓			✓		✓	✓					✓	
206	Xu et al. (2016)	✓						✓		✓	✓				
207	Yeniyurt et al. (2013)	✓					✓		✓	✓		✓			
208	Zaki et al. (2017)			✓				✓	✓				✓		

#	Reference											
209	Zangenehpour (2015)	✓									✓	
210	Zhan et al. (2017)	✓	✓			✓	✓				✓	
211	Zhang et al. (2015)			✓			✓			✓		
212	Zhao and Rosen (2017)		✓		✓			✓	✓			
213	Zhao et al. (2017)			✓		✓	✓			✓		
214	Zhong et al. (2013)		✓		✓	✓	✓	✓			✓	
215	Zhong et al. (2015a)	✓				✓				✓		
216	Zhong et al. (2015b)	✓				✓				✓		
217	Zhong et al. (2016)	✓			✓	✓	✓			✓		
218	Zhong et al. (2017)			✓	✓	✓	✓			✓		
219	Zhou et al. (2017)										✓	
220	Zhu (2014)	✓					✓			✓		

Table 3 Classification of papers on level of analytics used in different SCOR domains

SCOR domain	Descriptive	Predictive	Prescriptive
Plan	26	23	9
Source	7	9	8
Make	11	29	22
Deliver	13	32	14
OSC	18	21	12

Table 4 SCM resources utilized in different SCOR domains

SCOR domains	Financial	Physical	Human	Org	Tech	Int.
Plan	5	9	7	36	26	33
Source	3	3	4	16	13	5
Make	6	15	4	36	36	1
Deliver	3	30	5	23	26	13
OSC	3	4	5	23	25	5

Fig. 8 Distribution of papers in plan domain

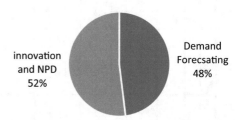

innovation and NPD 52%

Demand Forecsating 48%

financial, physical, human, organizational, technological and intangibles) and the SCOR domain (viz. plan, source, make and deliver). Any paper that did not indicate one or more particular practices was classified under 'Overall SC' or 'OSC'.

3.1 BDA Applications in Plan Domain

The studies in this domain are further classified into two categories managing innovation – new product development and demand forecasting. Among the five areas of SCOR domains, plan (58 out of 220 papers, 26%) is one of the most common areas where BDA supports decision-making. It is observed from Fig. 8 that 30 papers (or 51%) focus on using BDA for garnering innovation and new product development, while the remaining 28 papers (48 %) in this domain focus on using BDA for demand forecasting.

Figure 9 shows that descriptive (26 papers or 44%) and predictive (23 papers or 39%) analytics is widely used in the planning domain. Descriptive analytics is prominent in most studies on new product development and innovation. Out of the 28 papers on demand forecasting, 17 (or 60%) used predictive analytics, and nine (or 32%) used prescriptive approaches to BDA. The use of prescriptive analytics is primarily used in combination with predictive analytics.

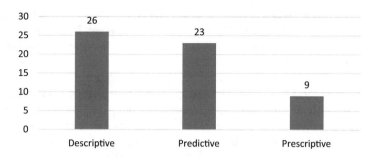

Fig. 9 Level of analytics in plan domain

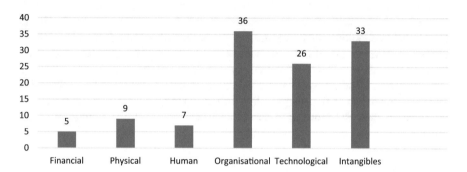

Fig. 10 Distribution of papers on SCM resources in plan domain

Further, as seen in Fig. 10, majority of the studies manage organizational resources (36 papers or 62%), intangible resources (33 papers or 56%) and technological resources (26 papers or 44%). The intangible resources get prominence in the plan domain because of the involvement of the customers and the users in giving their feedback for product development and the innovation process. Human and financial resources have not received much prominence in the plan domain.

3.1.1 New Product Development and Innovation

In online product reviews, BD is a major information source which helps managers and marketers realize their customers' concerns and interests (Xiao et al., 2015; Bendoly et al., 2012). Referred to as social and sentiment analysis, BDA helps design creative product strategies, launch new products to market as quickly as possible and determine the product's weaknesses earlier in the development cycle (Zhan et al., 2017; Lee & Chang, 2010; Colace et al., 2014). Tanev et al. (2015), based on their work, observed that the value of product-enabled services in an online environment is dependent on various SCM parameters. Therefore being in touch with the customers for their valuable feedback and input is highly important. The different sources of BD used for NPD are online shopping sites, blogs, social network sites and forums (Amarouche et al., 2015). A salesperson's knowledge of marketing intelligence is also an important source of business intelligence (Kuester & Rauch, 2016; Mariadoss et al., 2014). It is proposed to combine the lead user intelligence with the voice of the customer techniques to lower the risk of an unreliable product (Sann et al., 2013). Jin et al. (2016) recommend that product designers analyse BD in customer opinion data, purchase records and online behaviour using the Kalman filter and Bayesian method. Chung and Tseng (2012) suggest techniques for analysing the qualitative and quantitative data available on the web.

Studies have proposed that the concept of BD has substantial implications for R&D and innovation management (Blackburn et al., 2017; Bradley et al., 2017; Lau et al., 2014). BD and BDA are not only used extensively for NPD but also ensure product success (Xu et al., 2016; Ünay & Zehir, 2012; Bahrami et al., 2012; Eidizadeh et al., 2017; Schoenherr & Swink, 2015). Haverila and Ashill (2011) found that technology-intensive managers conceptualize and recognize 'intelligence' variables in successful and unsuccessful NPD projects. However, Tan et al. (2015) observed an absence of analytic data methods to support firms in capturing the innovation afforded by data and gaining a competitive advantage. Companies must develop a data analytic approach to utilize BD to gain a competitive advantage by boosting their SC innovation capabilities (Tan et al., 2015; Veugelers et al., 2010).

3.1.2   Demand Forecasting

BDA is used for obtaining an improved forecast, given that the company can identify the hidden trends and patterns from the data. Lamba and Singh (2017) supported that demand forecasting assists towards demand estimates. Process variations resulting from poor forecasts and demand predictions cause an imbalance leading to SC disruptions (Wang et al., 2016a; Souza, 2014; Chen & Blue, 2010; Chen et al., 2010). It is observed from the literature that the BD and BDA applications in demand forecasting are mainly for three purposes, viz. demand planning, demand sensing and demand shaping. Demand planning deals with analysing the different customer segments that help organizations create revenue plans (Chen & Blue, 2010; Haberleitner et al., 2010). Ulrike et al. (2013) proposed integrating sophisticated procedures to meet the data volume and complexity challenges. Automating analytics using adaptive forecasting time series benefits from performing forecasts of many variables with relatively high accuracy for a short period and few resources (Gerunov, 2016).

Demand sensing is a forecasting method performed on real-time information combined with new mathematical techniques to predict demand forecasts. With demand sensing and real-time analytical capabilities, organizations can analyse demand data with increasing real-time accuracy and reducing the bullwhip effect (Hofmann, 2017). Berengueres and Efimov (2014) used a linear extrapolation with GBM and other data mining techniques. The predicted information is recommended for CRM interactions between the airline and the passenger. Ma et al. (2014)

developed an algorithm for predicting future sales by capturing hidden and upcoming product demand trends. The proposed analytics was an integration of three prediction techniques, viz. (i) decision tree for large-scale data, (ii) discrete choice analysis for demand modelling and (iii) automatic time-series forecasting for trend analysis. Packianather et al. (2017) recommended using k-means clustering, hierarchical clustering and time-series forecasting to determine associations between customer variables identifying the seasonal variations and trends to visualize the core characteristics of the firm's customers. Cluster analysis and decision trees were used to classify traffic patterns to predict the electric vehicle charging demand (Arias & Bae, 2016).

Customer opinions captured from social media are used for competitive analytics and sentiment benchmarks. The findings of such studies are used for identifying specific, actionable areas (Chong et al., 2016; He et al., 2015; Marine-Roig & Clavé, 2015; Salehan & Kim, 2016). The BDA techniques used for customer analytics include predictive analytics such as sentimental and neural networks used on online reviews (Chong et al., 2016), collected from amazon.com (Li et al., 2016a), and search traffic information using Google Insights (Jun et al., 2014) and Twitter hashtags (Chae, 2015).

3.2   BDA Applications in Source Domain

The source domain has received very little interest from the researchers compared to the other SCOR domains (other than the return domain), with only 24 papers (or 16%). As seen in Fig. 11 use of BDA in the procurement process (ten papers, 41%) has received more attention in the source domain. In comparison, the remaining papers in this domain are well distributed between supplier selection (six papers, 25%), supplier performance (four papers, 16%) and supplier risk management (four papers or 16%).

It is observed from Fig. 12 that all three-level analytics are almost equally balanced in their application in the source domain, with predictive analytics (nine papers or 37%) having a little higher edge over the other analytics. Prescriptive analytics is primarily used in the source domain to detect frauds and minimize supplier risks.

Fig. 11 Distribution of papers in source domain

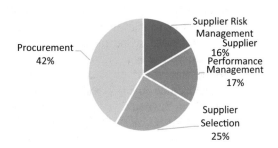

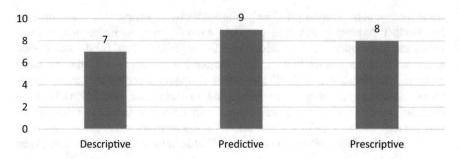

Fig. 12 Level of analytics in source domain

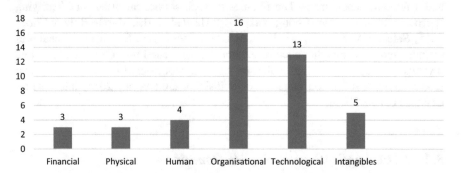

Fig. 13 Distribution of papers on SCM resources in source domain

Figure 13 reveals that most studies focused on managing the organizational resources (16 papers or 66%) as the BDA application serves to find hidden trends and fault detection in the procurement process, optimize the supplier selection process and minimize the risks. Technological resources were addressed in 13 papers (or 54%), followed by intangibles (five papers or 20%). The other resources, viz., financial (three papers or 12%), physical (three papers or 12%) and human (four papers or 16%), are least preferred in the literature.

3.2.1 Procurement

Many different types of data like spending details, supplier information, attribute criteria, etc. get generated during the procurement process from various sources. There is considerable scope for using BDA in procurement for beneficial info delivery (Westerski et al., 2015), understanding the procurement patterns of the customers (Mishra et al., 2016) and a variety of data-based analyses for business decisions that include quality problems and material availability (Souza, 2014; Min, 2010) not only in the traditional procurement system but also in the e-procurement environment (Wu et al., 2013). More and more enterprises are adopting e-procurement mode for purchase transactions, thereby generating a different combination of parameters increasing the difficulty level of the customer's choice.

Identifying this substantial potential, Choi et al. (2016) recommend using BDA to exploit the full potential of BD availability. With the help of a case study on sound public decision-making regarding IT service procurement, the authors demonstrated sound public decision-making. The value of BDA is not only derived from private companies but also from public procurement (Choi et al., 2016; Mirsolav et al., 2014). Detection of the fraud in public procurement has also been one of the areas of research in procurement that has received attention. BDA techniques like semantic technologies (Miroslav et al., 2014), supervised learning (Arief et al., 2016) and association rules (Ralha et al., 2012) were proposed for early recognition of potentially irregular procurement. The fraud detection studies are performed by extracting useful information from procurement process databases. Groves (2014) performed the simulation for a one-product life cycle with six autonomous agents competing to procure parts and sell the finished products to customers. The use of simulation provides insights applicable to SC environments based on prescriptive analytics.

3.2.2 Supplier Selection

A manufacturer or an assembler procures various raw materials, components and subassemblies from different suppliers based in different locations, producing the final product. In most cases, these companies have many suppliers to select from, offering them to supply the required products. Supplier selection, therefore, becomes a critical decision for managing an efficient supply chain. The companies use different criteria to select the best supplier(s). As the supplier selection decision must be optimized, the decision-making techniques use the knowledge and experience of the decision-makers (Jain et al., 2014). BDA in supplier selection helps in discovering the hidden relationships. The same can be achieved by analysing the supplier's pre-qualification data (Jain et al., 2014), firm profiles and transactional relationships (Mori et al., 2012) and historical data (AlKhalifah & Ansari, 2016) by using different techniques like association rule mining (Kuo et al., 2015; Lin et al., 2010), artificial intelligence, machine learning (Mori et al., 2012) and optimization techniques (Kuo et al., 2015). Two-stage supplier selection models with the first stage focusing on selecting the supplier and the second stage deciding on the quantity allocation for the key suppliers to minimize the purchasing cost are also recommended (Kuo et al., 2015; Lin et al., 2010).

3.3   BDA Applications in Make Domain

It is observed from Fig. 14 that the making domain takes up 28% (62 out of 220 papers) of the total publications. The majority of research papers in this area (36 papers, 58%) focus on using BDA for improving the manufacturing environment and achieving process improvements, followed by application in quality management (ten papers, 16%), maintenance management (seven papers, 11%) and scheduling and production control (seven papers, 11%).

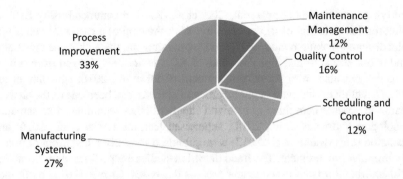

Fig. 14 Distribution of papers in make domain

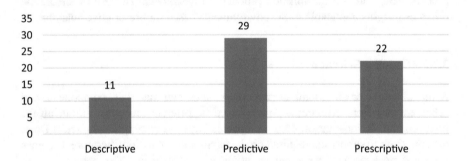

Fig. 15 Level of analytics in make domain

Figure 15 reveals that the level of analytics is more inclined towards predictive (29 papers or 46%) and prescriptive analytics (22 papers or 35%). Descriptive analytics is used to a lesser extent (11 papers or 17%).

Further, the focus of all the studies was to use the BDA to manage and use data for organizational decision-making. Therefore, we find that most studies dealt with managing the technological and organizational resources (36 papers each or 58% each). Physical resources were the focus of the study in 15 papers (or 24%). Managing the financial resources (six papers or 9%), human resources (three papers or 4%) and intangible resources (two papers or 3%) has received very little attention in the make domain (see Fig. 16).

3.3.1 Manufacturing Systems

There is increasing pressure on the manufacturing enterprises to improve their efficiency and productivity due to the increasing and changing customer demands. Many factors, including various complex products, uncertainties, capacity constraints and labour shortages, make production management face more and more significant challenges (Butler and Bright, 2014; Cheng et al., 2017; Zaki et al., 2017). Internet of Things (IoT) is the organic evolution of the Internet that creates a

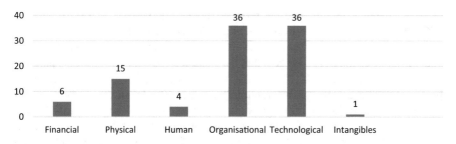

Fig. 16 Distribution of papers on SCM resources in make domain

smart environment (Mourtzis et al., 2016; Davis et al., 2012; Lade et al., 2017), giving scope for the intensive use of BDA (Babiceanu & Seker, 2016; Cheng et al., 2017; Lee et al., 2013). With the increasing connectivity of devices, the rapid growth of data recorded and ready for analysis is growing correspondingly (Zhu et al. 2014). Liu et al. (2016) found that customized manufacturing tasks could be finished more reliably and efficiently if all participants exchanged the data through a cloud platform in real time.

Prescriptive analytics, like multi-objective optimization, has been identified as a robust approach for generating a set of optimal trade-off design alternatives (Dudas et al., 2014; Munro & Madan, 2016.) in manufacturing. Few authors have proposed using a combination of computational intelligence approaches combining optimization, simulation and data mining to optimize the manufacturing systems (Amos et al., 2016, Jain et al., 2017). The use of simulation as a BDA technique in manufacturing is used as a tool for data generator and model validation. Kibira et al. (2015) recommend using BDA to extract significant parameters that affect the system performance, using these parameters as input values for simulation and then using the simulation output to optimize the system. Virtual manufacturing environments are proposed to be developed to capture, store, reuse and share manufacturing knowledge (Shafiq et al., 2017). The virtual environments are reported to provide for the collective intelligence of a factory and enhance effective decision-making (Kretschmer et al., 2017). Soroka et al. (2017) find that BDA in redistributed manufacturing (RdM) may help small-scale manufacturing companies to manufacture tailored products satisfying the specific needs of consumers (Zaki et al., 2017; Soroka et al., 2017).

3.3.2 Process Improvement

A typically processed lot in manufacturing generates huge data, which needs proper analysis for process troubleshooting (Cochran et al., 2016 Charaniya et al., 2010). Djatna and Munichputranto (2015) used overall equipment effectiveness as a quantitative productivity measurement for continuous improvement and evaluation. Using the Android-based mobile BI system, they identified the critical production line effectiveness measurement and machine utilization parameters. Apart from process analytics, manufacturing process modelling (Çiflikli & Özyirmidokuz, 2010; Kwak & Kim, 2012; Robinson et al., 2015) and process simulation (Soban et al., 2016) are also observed to be an area of interest for BDA application. Çiflikli and Özyirmidokuz (2010) suggest that the nonlinearities between process parameters, which are not inevitable in manufacturing processes, can also be addressed by process modelling. Energy consumption (Shin et al., 2014), energy costs (Hammer et al., 2017) and reduction of the greenhouse gas emissions (Wang et al., 2016c) are optimized using BDA for attaining sustainable manufacturing. The process optimization technique proposed by Kwak and Kim (2012) addresses the issue of handling a significant amount of missing values due to the data discarded by gross measurement errors. The IoT and wireless technologies like RFID found their application in the shop floor environment, referred to as intelligent shop floors (Zhong et al., 2017). Charaniya et al. (2010) developed a kernel-based methodology to integrate all the process parameters. Proposed using a genetic algorithm-support vector machines method to identify the input process variables for a cleaner production environment.

3.3.3   Maintenance Management

Predictive maintenance represents one area where BD solutions benefit various process types. Predictive analytics enables immediate data collection for analysis by the data aggregation and merging functions which extract keys correlating to yield from the equipment's parameter for detecting the root cause (Tsuda et al., 2015; Moyne et al., 2017). Wang et al. (2017) discussed predictive analytics where the schedules may be optimized using the linguistic interval-valued fuzzy reasoning method (Kumar et al., 2017). The decision tree model as a predictive mining technique was used for detecting and isolating machine breakdowns (Cochran et al., 2016).

3.3.4   Scheduling and Production Control

The exposure of the manufacturing companies to volatile market conditions causes wide variations and discrepancies in executing production plans and schedules. Using real-time shop floor data about men, machines, materials and orders captured through RFID technology can be an important source for developing the standard operating times on which the advanced production plan and schedules are prepared (Zhong et al., 2013). BDA has a huge potential for production planning and control (Krumeich et al. (2016), control engineering analyses and managing the dynamically changing production circumstances and huge varieties in production programmes (Reuter et al., 2016). Prescriptive analytics was used to optimize the scheduling problem, which is usually NP-hard (Cohen et al., 2017), and the performance of industrial manufacturing systems (Hammer et al., 2017) to maximize the enterprise profit.

3.4 BDA Applications in Deliver Domain

Another domain for BDA applications is delivered with 60 out of 220 papers (or 27%). Figure 17 shows that transport and traffic management (26 papers, 43%) and inventory management (16 papers or 26%) are the most prominent application areas of BDA in this domain. BDA use for logistics network management (ten papers, 16%) and order picking (eight papers, 13%) is also gaining importance in the recent past.

Predictive analytics has been the most popular technique, with 32 papers (or 53%) out of 62 papers in the deliver domain, followed by prescriptive analytics with 14 papers (or 23%). BDA using descriptive studies has attracted less attention from the researchers, with only 11 papers (or 21%) using them (see Fig. 18).

It is observed from Fig. 19 that the use of BDA for managing physical resources has been observed to find high visibility in the delivery domain, with the majority of the studies (30 papers or 50%) focusing on them. This was followed by studies also focusing on the issues of data collection (technological resources, 26 papers or 43%) and using the data for improving the organizational capabilities (23 papers, 38%). Managing the financial resources (3 papers or 5%), human resources (five papers or 8%) and intangible resources (13 papers or 21%) has received very little attention in the deliver domain.

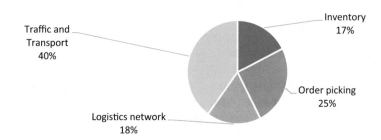

Fig. 17 Distribution of papers in deliver domain

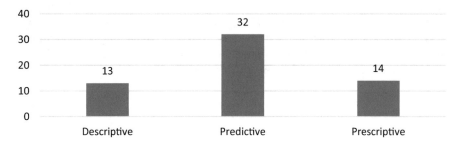

Fig. 18 Level of analytics in deliver domain

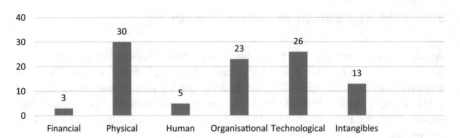

Fig. 19 Distribution of papers on SCM resources in make domain

3.4.1   SC Network Management

Park et al. (2016) implemented an interactive web-based visual analytic system to provide detailed information about supply network activities on demand management. The proposed system enables the decision-makers to interactively use visual encodings to match the demand gaps between the network nodes (Davis et al., 2012; Ilie-Zudor et al., 2015). Shukla and Kiridena (2016) developed an advanced analytics framework for configuring SC networks using historical sales data, including network node-related information. The primary focus of most of the studies on BDA in the supply network area has been sustainability and green SCM (Papadopoulos et al., 2017; Shukla & Kiridena, 2016; Zhao et al., 2017; Davis et al., 2012, Wu et al., 2017a, b). Wireless sensor network (WSN) technologies connecting to IoT and Automated Meter Reading (AMR) systems were used to collect the BD.

3.4.2   Order Picking

Many solution methodologies minimize the travelling distance for order picking (Chuang et al. (2014); Li et al. (2016b); Pang and Chan (2017); Azadnia et al. (2013)). Chuang et al. (2014) explored how an effective layout zoning following class-based storage can enhance order picking efficiency using association web statistics and association rule mining. Li et al. (2016b) also proposed improvement in order picking time in stores using an integrated mechanism based on the ABC categorization. BDA is used for the shelf space allocation strategy, leading to increased store profitability (Tsai & Huang, 2015) and reduced travel times (Chiang et al., 2014). The proposed solution captured customers' purchasing behaviour from records of previous transactions captured through RFID (Tsai & Huang, 2015) and analysed using association rule mining (Chiang et al., 2014).

3.4.3   Inventory Management

Maintaining an optimal inventory level in the organization is one of the most significant requirements for a competitive supply chain. Automation and integration of warehousing systems are inevitable for more efficient and centralized distribution (Alyahya et al., 2016). The various inventory control techniques and methodologies aim to reduce the overall SC costs by efficiently controlling the inventory. BD and BDA have been applied in this area to minimize the shortages and avoid overstocking of the products (Guo et al., 2014), as a variety of data in the form of historical demands, forecasts, etc., which are readily available to the organizations. Therefore, inventory management requires particular attention in the field of SCM. The BDA must provide accurate and up-to-date information for better user-friendly inventory management decisions (Lee et al., 2016a, b, c; Stefanovic, 2015). This is done by tracking the inventory levels, orders and sales by implementing intelligent inventory management solutions to reveal hidden relations with integrated data-driven analysis (Zhou et al., 2017), predicting inventory-level requirements based on average consumptions over a period and using machine learning algorithms (Thiruverahan and Subramanian, 2015), back-propagation neural network to train the prediction model (Guo et al., 2014; Hsu et al. (2015) and genetic algorithm (Thotappa and Ravindranath, 2010). BDA has a high potential for the improvement of the various SC processes that include mitigation of the bullwhip effect, demand variability at various nodes in the SC network and determination of optimum safety stock at various storage points in the SC (Guo et al., 2014). Integrated analytics based on operations research, data mining and geographic information can be used for in-transit inventory (Delen et al., 2011). Huang and van Meighem (2014) propose a dynamic decision support model that takes orders offline to reduce the inventory holding and back-ordering costs. Consideration of the green aspects of the SC for reducing the environmental risks due to the sediment storage, transport and deposition of BDA using satellite imagery and elevation data captured through GIS (Balaban et al., 2015) is also studied.

3.4.4   Transportation and Logistics

Transportation and logistics planning is an emerging field in which the organizations can have an improved business sense and SC decisions by taking advantage of BD enabled by the popularization of intelligent transportation systems (Shi and Abdel-Aty, 2015; Wang et al., 2016b; Kemp et al., 2016; Kubac, 2016). Many moving objects in the supply chain can be traced and tracked using precise transportation modes to model the BD (Xu & Güting, 2013). The research studies have focused more on developing a smart logistics environment with an ITS with the objective of prompt delivery aligning just-in-time standards (Dobre & Xhafa, 2014); Sivamani et al., 2014; Miyaji 2015).

The sensed data needs formatting and standardizing for further deployment (Zhong et al., 2015a, b). As one of the data collection sources, RFID technology helps capture real-time data used for different purposes, including keeping the inventory records, as any inaccuracies in these records will affect the SC performance (Kok & Shang, 2014). The collected data can be temperature and humidity data during storage, transportation, event files and geographic and demographic data for efficient trip management (Cristobal et al., 2015). Human or organizational factor data clubbed with real-time traffic and weather data were used to predict truck arrival times (van der Spoel et al., 2017). Ergonomic methods applied to BD collected from On-Train Data Recorders (OTDR) can help address the risk issues and improve human performance (Walker & Strathie, 2016). Wallander & Makitalo (2012) used passenger train traffic data to analyse the transport delay chains, which can be used to develop rail traffic punctuality and the whole railway system to improve the rail network system.

Cui et al. (2016) used GPS devices as a data source to minimize the difference between travel demand and transport services, leading to a sustainable urban transport system. Critical traffic data were captured by the devices installed on the vehicles. Many research studies felt the need to integrate new BD resources into customary transportation demand modelling so that the increasing stress on already burdened transportation infrastructure, waiting time, congestion and accidents happening due to rapid urbanization are reduced through proactive real-time traffic monitoring (Toole et al., 2015; Cristobal et al., 2015; Xiao et al., 2015; Zangenehpour et al., 2015; Shi & Abdel-Aty, 2015; Wang et al., 2016b).

It has been found that transitionary technologies are presently leading the way in capturing the wealth of information for intelligent transportation systems (Shan Zhu, 2015; Zangenehpour et al., 2015). St.-Aubin et al. (2015) provide a functional framework for implementing an automated, high-resolution, video-based traffic-analysis system to conduct a road safety analysis and validate traffic flow models. Another significant application of BDA techniques that were observed in the literature included techniques such as granular computing (Xie, 2016), multiresolution data aggregation and visual data mining (Wang et al., 2016c), association rule mining (Lanka & Jena, 2014; Diana, 2012; Kargari & Sepehri (2012)), neural networks (Li et al., 2014) and network routing, scheduling and real-time control algorithms (Levner et al., 2011). Kargari and Sepehri (2012) suggested clustering retail stores in a distribution network considering the available information to reduce distribution and transportation costs, such as store location, order, goods, vehicles and road and traffic information.

To select a route according to the user's requirement, the model uses necessary delivery information, including location, delivery vehicle, user, etc. With the increased Internet usage, companies can gather real-time situations to develop a more reliable relationship as an outcome of the Internet of Things. Sustainable and green transportation systems have also been the focus of the research studies (Zhao et al., 2017; Lee et al., 2017; Tu et al., 2016). Ehmke et al. (2016) introduced the shortest path algorithm that incorporated flexibility in travel speeds and estimated arrival time distributions at nodes on a path to reduce emissions relative to minimum distance and time-dependent paths. Tu et al. (2016) addressed the issue of the absence of charging stations as a limiting factor for the penetration of electric vehicles. Mehmood (2017) studied the transport operations to deal with lowering CO2 footprint (O'Brien et al., 2014). However, most models developed are deterministic (Petri et al., 2016).

4   SC Visibility, BDA Capability and SC Transformation

The findings of the SLR indicate that SC decision-makers are required to process voluminous data for decision-making to reduce SC costs and increase product availability, meeting the customer demands. For providing the required assistance to the decision-makers with rich and quality data, they must be available to the organization along with organizational infrastructure in terms of technological resources, physical systems and processes, human expertise and financial resources for collection of data, storage of data and performing the data analysis for extracting meaningful information leading to successful supply chain transformation.

4.1   SC Visibility Framework

The SLR reveals that BDA provides an excellent opportunity for strategic and operational improvements in the supply chains (Wang et al., 2016c). However, firms must be able to grab these opportunities and convert them into actions (Srinivasan & Swink, 2017). In SCM, the managers must collect and analyse information from stakeholders across the SCOR dimensions discussed in this chapter to drive better decision-making. Our proposed framework uses SC visibility (Williams et al., 2013) as the primary driving force that leads to successful SC transformation through a robust BDA capability, as shown in Fig. 20.

4.1.1   SC Visibility

Srinivasan and Swink (2017) define SC visibility as the availability of appropriate and precise data from external partners, enabling the firms to build systems to process and acquire insights and synchronized decision-making between the SC partners. The SC visibility is required across all the domains of SC, viz. plan, source, make, deliver and return. The demand data enables the firms to sense the changing customer patterns, competitor actions, promotion actions, pricing strategies, demand forecasting, delivery planning, inventory decisions, NPD, etc. The supply data generated from the suppliers enables the firms to recognize the changes, costs, shipping notifications, delivery schedules, managing inventory, etc.

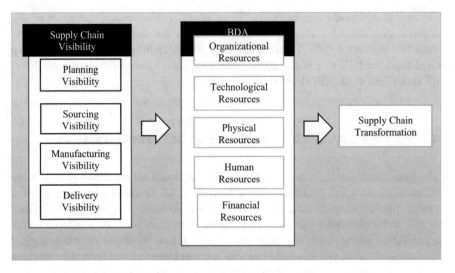

Fig. 20 SC visibility – BDA capability framework

4.1.2 BDA Capability

In this study, we conceptualize the BDA on the organizational facility with techno-logical tools and techniques, physical systems, human analytical skills and financial resources that enable the firms to collect, process, organize, visualize and analyse the data and use the derived information for enabling a big-data-driven efficient supply chain.

4.1.3 SC Transformation

Usually, a firm investing in achieving high SC visibility will also ensure well-developed resources that build their BDA capability. SC visibility will be useless if the firm has not developed its BDA capability or vice versa. To achieve an efficient SC, SC visibility and BDA capability must co-exist. The availability of rich infor-mation for operational and strategic decision-making will help the SC managers to reduce operational costs and improve product delivery performance.

5 Future Research Directions

The future directions to capitalize on the research development of BDA applications in the SCM context are discussed in this section.

5.1 Future Investigations on BDA Applications in the SCOR Domains

5.1.1 Plan Domain

BDA in NPD helps to reduce the risk and market uncertainties as the data is acquired through different sources at the early phases of product development (Wamba et al., 2015), which can be used to identify the previously unrecognized customer needs (Tsai & Huang, 2015), for generating new product ideas (Trkman et al., 2010), in the process developing a long-term relationship with customers acting as co-creators of the product lending customer loyalty and retention (Gantz & Reinsel, 2012). It is found from the literature that most studies on NPD and innovation focus on the sources of BD and different approaches for analysing the data. Further studies should be undertaken on improving customer involvement by using BD and organizations' cooperation with the customers and their involvement in the NPD process (Zhan et al., 2017).

5.1.2 Source Domain

The review suggests that the source domain has received the least attention from the researchers. More studies in this domain on different aspects of BDA applications in procurement modes, supplier selection and evaluation, supplier risk management, supplier contracts management, etc. must be taken in the future.

5.1.3 Make Domain

In the make domain, the Internet of Things (IoT) and smart manufacturing have started gaining importance and are expected to bring voluminous data regarding quantity and category, providing an opportunity for the application and development of BD. It is believed that the BDA in the IoT space will accelerate research advancements and business models (Min, 2010). BDA enables data set and comprehensive assessment from different sources and customers towards decision-making, improving customer service and manufacturing flexibility, optimizing production quality, saving energy and improving equipment service (Ilie-Zudor et al., 2015). Presently BDA in the make domain is used for process monitoring, fault finding and process optimizations, mainly based on the predictive and prescriptive analytics (Moyne et al., 2017). The real challenge in the future will be implementing BDA in smart manufacturing systems. Therefore the experts in the make domain must identify the critical tasks/processes for a more manageable outcome. Further, these IoT-based smart manufacturing systems are connected with embedded sensors and communication devices.

5.1.4 Deliver Domain

Supply chains have become more reliable, efficient and predictable with digital technologies in transportation and warehousing functions. The review reveals that the supply chains are accessing real-time data and analytics with the extended use of RFID and other IoT sensors that provide a live view of their machinery, vehicles and operators. Future studies should focus on ensuring a safe and pleasant working environment for the drivers and reducing traffic accidents and fatalities. Future studies may also focus on the impact of BDA on the potential use of autonomous vehicles. Studies on exploring how 3D printing can help to eliminate the transportation costs as it is expected that the 3D printing will help the manufacturing facilities to be moving closer to the customers hence reducing the distance and also the need for stocking inventories as the products can be manufactured whenever the demand is received. Further studies should focus on how the supply chains can adopt or adapt to new business models to stay on course, compete in today's digital market and ultimately embrace the transformative capabilities of BD for the industry.

The other research gaps which are required to be addressed in the future studies on BDA applications in deliver domain include in-plant logistics movement of raw materials, products and vehicles, just-in-time inventory management with the use of IoT and smart manufacturing systems, use of BDA for efficient use of robots in the warehouse and order picking management.

5.1.5 Return Domain

The review did not identify any significant applications of BDA in the return domain. The results are in sync with the past studies (Nguyen et al., 2017; Barbosa et al., 2017). There is a high research potential to fill this gap in future studies. Further studies may focus on how the companies may learn from customer complaints while returning a defective product in the supply chain, identify root causes for defects, use predictive analytics to predict return rates and conduct assessments of the return process performance. Presently, the research in the return domain may be lacking because of difficulty obtaining the field's information. Still, new technologies such as IoT and intelligent systems will soon overcome this barrier.

5.2 Levels of Analytics Across the SCOR Domains

In the plan domain, the descriptive and predictive level dominates over the prescriptive level of analytics. In the source domain, the level of analytics seems to be equally balanced within the limited number of articles available in this domain. In make domain, the studies are dominated by predictive and prescriptive analytics. Predictive analytics dominates the deliver domain, focusing less on deploying descriptive and prescriptive analytics. Nguyen et al. (2017) suggest that BDA application is a linear process to catalyze rapid progression. Future studies should balance the emphasis on all three levels of analytics. The findings of this review will be useful guidance for future studies on identifying the relevant level of analytics in a given domain.

5.3   SCM Resources

Most studies on BDA applications in SCM focus on utilizing organizational and technological resources. Studies utilizing the intangible resources for managing NPD and product innovations were observed in the plan domain, and studies utilizing the physical resources were observed in deliver and make domain. Very few studies focused on the utilization of financial and human resources. The BDA capabilities of organizations depend on how best the various resources are utilized within the supply chains. It is evident from the review that although enough attention has been given to the technological resources in the literature, the future studies will still have to address the challenges of utilizing the technological resources due to the increased generation of real-time data with the use of the IoT devices, sensors and embedded technology. More studies will be required to address the challenges of utilizing human resources. With the increased complexities in the data, there is a considerable demand for experts to handle the technology and perform the BDA (Chen et al., 2012; Richey et al., 2016; Schoenherr & Speier-Pero 2015; Barbosa et al., 2017). These skills may be categorized as technical skills for collecting, storing and retrieving data and analytical skills for analysing the qualitative and quantitative data generated from different domains.

6   Conclusions and Limitations

This chapter offers a holistic view of BDA applications in the SC context. These SLR's findings will guide academicians and practitioners working in the BDA area to build for such challenges. Based on the study results, we propose an SC visibility framework that identifies SC visibility as the main driving force for successful SC transformation, achieved through strong BDA capability. The findings of this SLR and future research directions will help the academics, researchers and practitioners to focus on the BDA challenges. The authors recognize that our study has limitations. While the authors have conducted a thorough literature search through different research databases like Scopus, Web of Science, Emerald Insight, etc., to identify all possible relevant articles, it is possible that some research articles could have been missed in this review and may be further explored.

References

Addo-Tenkorang, R., & Helo, P. T. (2016). Big data applications in operations/supply-chain management: A literature review. *Computers & Industrial Engineering, 101*, 528–543.

Ahiaga-Dagbui, D. D., & Smith, S. D. (2014). Dealing with construction cost overruns using data mining. *Construction Management and Economics, 32*(7–8), 682–694.

Akter, S., Wamba, S. F., Gunasekaran, A., Dubey, R., & Childe, S. J. (2016). How to improve firm performance using big data analytics capability and business strategy alignment? *International Journal of Production Economics, 182*, 113–131. https://doi.org/10.1016/j.ijpe.2016.08.018

AlKhalifah, A., & Ansari, G. A. (2016). *Modeling of E-procurement system through UML using data mining technique for supplier performance.* In International Conference on Software Networking (ICSN) (pp. 65–70).

Aloysius, J. A., Hoehle, H., Goodarzi, S., & Venkatesh, V. (2016). Big data initiatives in retail environments: Linking service process perceptions to shopping outcomes. *Annals of Operations Research, 270*, 25–51.

Alyahya, S., Wang, Q., & Bennett, N. (2016). Application and integration of an RFID-enabled warehousing management system–a feasibility study. Journal of Industrial Information Integration, *4*, 15–25.

Amarouche, K., Benbrahim, H., & Kassou, I. (2015). Product opinion mining for competitive intelligence. *Procedia Computer Science, 73*, 358–365.

Amos, H. C. N., Bandaru, S., & Frantzén, M. (2016). Innovative design and analysis of production systems by multi-objective optimization and data mining. *Procedia CIRP, 50*, 665–671.

Arias, M. B., & Bae, S. (2016). Electric vehicle charging demand forecasting model based on big data technologies. *Applied Energy, 183*, 327–339.

Arief, H., Saptawati, G. A. P, & Asnar, Y. D. W. (2016). *Fraud detection based-on data mining on Indonesian e-procurement system.* In International Conference on Data and Software Engineering (ICoDSE) (pp. 176–182).

Arunachalam, D., Kumar, N., & Kawalek, J. P. (2017). Understanding big data analytics capabilities in supply chain management: Unravelling the issues, challenges and implications for practice. *Transportation Research, 114*, 416–436.

Assunção, M. D., Calheiros, R. N., Bianchi, S., Netto, M. A., & Buyya, R. (2015). Big data computing and clouds: Trends and future directions. *Journal of Parallel and Distributed Computing, 79–80*, 3–15.

Azadnia, A. H., Taheri, S., Ghadimi, P., Saman, M. Z. M., & Wong, K. Y. (2013). Order batching in warehouses by minimizing total tardiness: A hybrid, approach of weighted association rule mining and genetic algorithms. *Scientific World Journal, 2013*, 1–13.

Babiceanu, R. F., & Seker, R. (2016). Big data and virtualization for manufacturing cyber-physical systems: A survey of the current status and future outlook. *Computers in Industry, 81*, 128–137.

Bag, S. (2016). Fuzzy VIKOR approach for selection of big data analyst in procurement management. *Journal of Transport and Supply Chain Management, 10*(1), 1–6.

Bahrami, S. M., Arabzad, M., & Ghorbani, M. (2012). Innovation in market management by utilizing business intelligence: Introducing proposed framework. *Procedia-Social and Behavioral Sciences, 41*, 160–167.

Balaban, S. I., Hudson-Edwards, K. A., & Miller, J. R. (2015). A GIS-based method for evaluating sediment storage and transport in large mining-affected river systems. *Environmental Earth Sciences, 74*(6), 4685–4698.

Barbosa, M. W., Vicente, A., Ladeira, M. B., & Oliveira, M. P. (2017). Managing supply chain resources with big data analytics: A systematic review. *International Journal of Logistics Research and Applications*, 1–24.

Barratt, M., & Barratt, R. (2011). Exploring internal and external supply chain linkages: Evidence from the field. *Journal of Operations Management, 29*(5), 514–528.

Bauer, C., Siddiqui, Z. F., Beuttler, M., & Bauer, K. (2016). Big Data in manufacturing systems engineering – Close up on a machine tool. *Automatisierungstechnik, 64*(7), 534–539.

Bendoly, E. (2016). Fit, bias, and enacted sense making in data visualization: Frameworks for continuous development in operations and supply chain management analytics. *Journal of Business Logistics, 37*(1), 6–17.

Bendoly, E., Bharadwaj, A., & Bharadwaj, S. (2012). Complementary drivers of new product development performance: Cross-functional coordination, information system capability, and intelligence quality. *Production and Operations Management, 21*(4), 653–667.

Berengueres, J., & Efimov, D. (2014). Airline new customer tier level forecasting for real-time resource allocation of a miles program. *Journal of Big Data, 1*(1), 3–5.

Bhattacharjya, J., Ellison, A., & Tripathi, S. (2016). An exploration of logistics-related customer service provision on twitter: The case of e-retailers. *International Journal of Physical Distribution and Logistics Management, 46*(6/7), 659–680.

Blackburn, M., Alexander, J., Legan, J. D., & Klabjan, D. (2017). Big data and the future of R and D management. *Research-Technology Management, 60*(5), 43–51.

Bradley, R., Jawahir, I. S., Murrell, N., & Whitney, J. (2017). Parallel design of a product and Internet of Things (IoT) architecture to minimize the cost of utilizing Big Data (BD) for sustainable value creation. *Procedia CIRP, 61*, 58–62.

Braganza, A., Brooks, L., Nepelski, D., Ali, M., & Moro, R. (2017). Resource management in big data initiatives: Processes and dynamic capabilities. *Journal of Business Research, 70*, 328–337.

Brandenburger, J., Colla, V., Nastasi, G., Ferro, F., Schirm, C., & Melcher, J. (2016). Big data solution for quality monitoring and improvement on flat steel production. *IFAC-PapersOnLine, 49*(20), 55–60.

Brinch, M., Stentoft, J., & Jensen, J. K. (2017). *Big data and its applications in supply chain management: Findings from a Delphi study.* In Proceedings of the 50th Hawaii international conference on system sciences. DOI: https://doi.org/10.24251/HICSS.2017.161

Butler, L. J., & Bright, G. (2014). Computational intelligence for advanced manufacturing system management: a review. *International Journal of Intelligent Systems Technologies and Applications, 13*(4), 258–266.

Cárdenas-Benítez, N., Aquino-Santos, R., Magaña-Espinoza, P., Aguilar-Velazco, J., Edwards-Block, A., & Medina Cass, A. (2016). Traffic congestion detection system through connected vehicles and big data. *Sensors, 16*(5), 599–612.

Chae, B. K. (2015). Insights from hashtag supply chain and Twitter Analytics: Considering Twitter and Twitter data for supply chain practice and research. *International Journal of Production Economics, 165*, 247–259.

Chae, B., & Olson, D. L. (2013). Business analytics for supply chain: A dynamic-capabilities framework. *International Journal of Information Technology & Decision Making, 12*(1), 9–26. https://doi.org/10.1142/S0219622013500016

Chae, B., Olson, D., & Sheu, C. (2014). The impact of supply chain analytics on operational performance: A resource-based view. *International Journal of Production Research, 52*(16), 4695–4710.

Charaniya, S., Le, H., Rangwala, H., Mills, K., Johnson, K., Karypis, G., & Hu, W. S. (2010). Mining manufacturing data for discovery of high productivity process characteristics. *Journal of Biotechnology, 147*(3–4), 186–197.

Chen, A., & Blue, J. (2010). Performance analysis of demand planning approaches for aggregating, forecasting and disaggregating interrelated demands. *International Journal of Production Economics, 128*(2), 586–602.

Chen, C., Ervolina, T., Harrison, T. P., & Gupta, B. (2010). Sales and operations planning in systems with order configuration uncertainty. *European Journal of Operational Research, 205*(3), 604–614.

Chen, H., Chiang, R. H. L., & Storey, V. C. (2012). Business intelligence and analytics: From big data to big impact. *MIS Quarterly, 36*(4), 1165–1188. https://doi.org/10.1145/2463676.2463712

Chen, Y. J., Chu-Yuan, F., & Chang, K. H. (2016). Manufacturing intelligence for reducing false alarm of defect classification by integrating similarity matching approach in CMOS image sensor manufacturing. *Computers and Industrial Engineering, 99*, 465–473.

Cheng, Y., Chen, K., Sun, H., Zhang, Y., & Tao, F. (2017). Data and knowledge mining with big data towards smart production. *Journal of Industrial Information Integration, 9*, 1–13.

Chiang, D. M. H., Lin, C. P., & Chen, M. C. (2011). The adaptive approach for storage assignment by mining data of warehouse management system for distribution centres. *Enterprise Information Systems, 5*(2), 219–234.

Chiang, M. H., Lin, C. P., & Chen, M. C. (2014). Data mining based storage assignment heuristics for travel distance reduction. *Expert Systems, 31*, 81–90.

Chien, C. F., Hsu, S. C., & Chen, Y. J. (2013). A system for online detection and classification of wafer bin map defect patterns for manufacturing intelligence. *International Journal of Production Research, 51*(8), 2324–2338.

Chien, C. F., Liu, C. W., & Chuang, S. C. (2017). Analysing semiconductor manufacturing big data for root cause detection of excursion for yield enhancement. *International Journal of Production Research, 55*(17), 5095–5107.

Choi, T. M. (2016). Incorporating social media observations and bounded rationality into fashion quick response supply chains in the big data era. *Transportation Research Part E: Logistics and Transportation Review, 114*, 386–397.

Choi, Y., Lee, H., & Irani, Z. (2016). Big data-driven fuzzy cognitive map for prioritizing IT service procurement in the public sector. *Annals of Operations Research, 243*(1–2), 1–30.

Chong, A. Y. L., Li, B., Ngai, E. W. T., Ch'ng, E., & Lee, F. (2016). Predicting online product sales via online reviews, sentiments, and promotion strategies: A big data architecture and neural network approach. *International Journal of Operations and Production Management, 36*(4), 358–383.

Chuang, Y. F., Chia, S. H., & Wong, J. Y. (2014). Enhancing order-picking efficiency through data mining and assignment approaches. *WSEAS Transactions on Business and Economics, 11*(1), 52–64.

Chung, W., & Tseng, T. L. (2012). Discovering business intelligence from online product reviews: A rule-induction framework. *Expert Systems with Applications, 39*(15), 11870–11879.

Çiflikli, C., & Özyirmidokuz, E. K. (2010). Implementing a data mining solution for enhancing carpet manufacturing productivity. *Knowledge-Based Systems, 23*(8), 783–788.

Cochran, D. S., Kinard, D., & Zhuming, B. (2016). Manufacturing system design meets big data analytics for continuous improvement. *Procedia CIRP, 50*, 647–652.

Cohen, M. W., Mitnovizky, M., & Shpitalni, M. (2017). Manufacturing systems: Using agents with local intelligence to maximize factory profit. *Journal of Manufacturing Science and Technology, 18*, 135–144.

Colace, F., Santo, M. D., & Greco, L. (2014). An adaptive product configurator based on slow intelligence approach. *International Journal of Metadata, Semantics and Ontologies, 9*(2), 128–137.

Cosic, R., Shanks, G., & Maynard, S. B. (2015). A business analytics capability framework. *Australasian Journal of Information Systems, 19*, No.5–19.

Cristobal, T., Lorenzo, J. J., & Garcia, C. R. (2015). *Using data mining to improve the public transport in Gran Canaria Island. Computer aided systems theory – Eurocast.* In International conference on computer aided systems theory (pp. 781–788).

Cui, J., Liu, F., Hu, J., Janssen, D., Wets, G., & Cools, M. (2016). Identifying mismatch between urban travel demand and transport network services using GPS data: A case study in the fast growing Chinese city of Harbin. *Neurocomputing, 181*, 4–18.

Davis, J., Edgar, T., Porter, J., Bernaden, J., & Sarli, M. (2012). Smart manufacturing, manufacturing intelligence and demand-dynamic performance. *Computers and Chemical Engineering, 47*, 145–156.

Delen, D., & Demirkan, H. (2013). Data, information and analytics as services. *Decision Support Systems, 55*(1), 359–363.

Delen, D., Erraguntla, M., Mayer, R. J., & Wu, C.-N. (2011). Better management of blood supply-chain with GIS-based analytics. *Annals of Operations Research, 185*(1), 181–193.

Delipinar, G. E., & Kocaoglu, B. (2016). Using SCOR model to gain competitive advantage: A literature review. *Procedia-Social and Behavioral Sciences, 229*, 398–406.

Diana, M. (2012). Studying patterns of use of transport modes through data mining application to us national household travel survey data set. *Transportation Research Record, 2308*, 1–9.

Dietrich, B., Ettl, M., Lederman, R. D., & Petrik, M. (2012). Optimizing the end-to-end value chain through demand shaping and advanced customer analytics. *Computer Aided Chemical Engineering, 31*, 8–18.

Djatna, T., & Munichputranto, F. (2015). An analysis and design of mobile business intelligence system for productivity measurement and evaluation in tire curing production line. *Procedia Manufacturing, 4*, 438–444.

Dobre, C., & Xhafa, F. (2014). Intelligent services for big data science. *Future Generation Computer Systems, 37*, 267–281.

Dubey, R. D., Gunasekaran, A., Childe, S. J., Wamba, S. F., & Papadopoulos, T. (2016). The impact of big data on world-class sustainable manufacturing. *International Journal of Advanced Manufacturing Technology, 84*(1-4), 631–645.

Dudas, C., Amos, H. C. N., Pehrsson, L., & Boström, H. (2014). Integration of data mining and multi-objective optimization for decision support in production systems development. *International Journal of Computer Integrated Manufacturing, 27*(9), 824–839.

Durán, O., Rodriguez, N., Consalter, L., & A. (2010). Collaborative particle swarm optimization with a data mining technique for manufacturing cell design. *Expert Systems with Applications, 37*(2), 1563–1567.

Ehmke, J. F., Campbell, A. M., & Thomas, B. W. (2016). Data-driven approaches for emissions-minimized paths in urban areas. *Computers & Operations Research, 67*, 34–47.

Eidizadeh, R., Salehzadeh, R., & Esfahani, A. C. (2017). Analysing the role of business intelligence, knowledge sharing and organisational innovation on gaining competitive advantage. *Journal of Workplace Learning, 29*(4), 250–267.

Emani, C. K., Cullot, N., & Nicolle, C. (2015). Understandable big data: A survey. *Computer Science Review, 17*, 70–81.

Fiosina, J., Fiosins, M., & Müller, J. P. (2013). Big data processing and mining for next generation intelligent transportation systems. *Journal Teknologi, 63*(3), 21–38.

Galbraith, J. R. (1973). *Designing complex organisations*. Addison-Wesley Longman Publishing Co., Inc..

Gandomi, A., & Haider, M. (2015). Beyond the hype: big data concepts, methods, and analytics. *International Journal of Information Management, 35*(2), 137–144.

Gantz, J., & Reinsel, D. (2012). The digital universe in 2020: Big data, bigger digital shadows, and biggest growth in the Far East. *IDC iView: IDC Analyze the Future, 2007*, 1–16.

Gao, D., Xu, Z., Ruan, Y. Z., & Lu, H. (2016). From a systematic literature review to integrated definition for sustainable supply chain innovation (SSCI). *Journal of Cleaner Production, 142*(4), 1518–1538.

Gerunov, A. (2016). Automating analytics: Forecasting time series in economics and business. *Journal of Economics and Political Economy, 3*(2), 340–349.

Govindan, K., Soleimani, H., & Kannan, D. (2014). Reverse logistics and closed-loop supply chain: a comprehensive review to explore the future. *European Journal of Operational Research, 240*(3), 603–626.

Groves, W., Collins, J., Gini, M., & Ketter, W. (2014). Agent-assisted supply chain management: Analysis and lessons learned. *Decision Support Systems, 57*, 274–284. https://doi.org/10.1016/j.dss.2013.09.006

Guo, X. X., Liu, C., Xu, W., Yuan, H., & Wang, M. M. (2014). *A prediction-based inventory optimization using data mining models*. In international joint conference on Computational Sciences and Optimization (CSO) (pp. 611–615).

Gupta, M., & George, J. F. (2016). Toward the development of a big data analytics capability. *Information and Management, 53*(8), 1049–1064.

Haberleitner, H., Meyr, H., & Taudes, A. (2010). Implementation of a demand planning system using advance order information. *International Journal of Production Economics, 128*(2), 518–526.

Hammer, M., Somers, K., Karre, H., & Ramsauer, C. (2017). Profit per hour as a target process control parameter for manufacturing systems enabled by big data analytics and industry 4.0 infrastructure. *Procedia CIRP, 63*, 715–720.

Haverila, M., & Ashill, N. (2011). Market intelligence and NPD success: a study of technology intensive companies in Finland. *Marketing Intelligence and Planning, 29*(5), 556–576.

Hazen, B. T., Boone, C., & A., Ezell, E, Z. and Farmer, A. J. (2014). Data quality for data science, predictive analytics, and big data in supply chain management: An introduction to the problem and suggestions for research and applications. *International Journal of Production Economics, 154*, 72–80.

Hazen, B. T., Skipper, J. B., Boone, C. A., & Hill, R. R. (2016). Back in business: Operations research in support of big data analytics for operations and supply chain management. *Annals of Operations Research*, 1–11.

He, W., Wu, H., Yan, G., Akula, V., & Shen, J. (2015). A novel social media competitive analytics framework with sentiment benchmarks. *Information Management, 52*(7), 801–812.

Ho, C. L., & Shih, H. W. (2014). Applying data mining to develop a warning system of procurement in construction. *International Journal of Future Computer and Communication, 3*(3), 168–171.

Hofmann, E. (2017). Big data and supply chain decisions: the impact of volume, variety and velocity properties on the bullwhip effect. *International Journal of Production Research, 55*(17), 5108–5126.

Hsu, C. Y., Yang, C. S., Yu, L. C., Lin, C. F., Yao, H. H., Chen, D. Y., Lai, R. K., & Chang, P. C. (2015). Development of a cloud-based service framework for energy conservation in a sustainable intelligent transportation system. *International Journal of Production Economics, 164*, 454–461.

Huang, T., & Van Mieghem, J. A. (2014). Clickstream data and inventory management: Model and empirical analysis. *Production and Operations Management, 23*(3), 333–347.

Huang, S. H., Sheoran, S. K., & Keskar, H. (2005). Computer-assisted supply chain configuration based on supply chain operations reference (SCOR) model. *Computers and Industrial Engineering, 48*(2), 377–394.

Ilie-Zudor, E., Ekárt, A., Kemeny, Z., Buckingham, C., Welch, P., & Monostori, L. (2015). Advanced predictive-analysis-based decision support for collaborative logistics networks. *Supply Chain Management: An International Journal, 20*(4), 369–388.

Ivanov, D. (2017). Simulation-based single vs. dual sourcing analysis in the supply chain with consideration of capacity disruptions, big data and demand patterns. *International Journal of Integrated Supply Management, 11*(1), 24–43.

Jain, R., Singh, A. R., Yadav, H. C., & Mishra, P. K. (2014). Using data mining synergies for evaluating criteria at pre-qualification stage of supplier selection. *Journal of Intelligent Manufacturing, 25*(1), 165–175.

Jain, S., Shao, G., Shin, S., & J. (2017). Manufacturing data analytics using a virtual factory representation. *International Journal of Production Research, 55*(18), 5450–5464.

Jeeva, A. S., & Dickie, C. (2012). A taxonomic approach to supplier intelligence in manufacturing: managing components of strategic procurement planning. *International Journal of Business Environment, 5*(1), 88–100.

Jelena, F., & Fiosins, M. (2017). Distributed Nonparametric and Semiparametric Regression on SPARK for Big Data Forecasting. *Applied Computational Intelligence and Soft Computing, 2017*, 1–13.

Jeon, S., & Hong, B. (2016). Monte Carlo simulation-based traffic speed forecasting using historical big data. *Future Generation Computer Systems, 65*, 182–195.

Ji-fan Ren, S., Fosso Wamba, S., Akter, S., Dubey, R., & Childe, S. J. (2017). Modelling quality dynamics, business value and firm performance in a big data analytics environment. *International Journal of Production Research, 55*(17), 5011–5026.

Jin, J., Liu, Y., Ji, P., & Liu, H. (2016). Understanding big consumer opinion data for market-driven product design. *International Journal of Production Research, 54*(10), 3019–3041.

Jun, S., Park, D., & Yeom, J. (2014). The possibility of using search traffic information to explore consumer product attitudes and forecast consumer preference. *Technological Forecasting and Social Change, 86*, 237–253.

Kache, F., & Seuring, S. (2017). Challenges and opportunities of digital information at the intersection of Big Data Analytics and supply chain management. *International Journal of Operations & Production Management, 37*(1), 10–36.

Kargari, M., & Sepehri, M. M. (2012). Stores clustering using a data mining approach for distributing automotive spare-parts to reduce transportation costs. *Expert Systems with Applications, 39*(5), 4740–4748.

Kemp, G., Solar, V. S., Da Silva, C. F., Ghodous, P., Collet, C., & Amalya, P. P. L. (2016). Cloud big data application for transport. *International Journal of Agile Systems and Management, 9*(3), 232–250.

Kibira, D., Qais, H., Soundar, K., & Guodong, S. (2015). *Integrating data analytics and simulation methods to support manufacturing decision making, Winter Simulation Conference (WSC).* In Winter simulation conference proceedings (pp. 2100–2111).

Kok, G., & Shang, K. H. (2014). Evaluation of cycle-count policies for supply chains with inventory inaccuracy and implications on RFID investments. *European Journal of Operational Research, 237*(1), 91–105.

Köksal, G., Batmaz, I., & Testik, M. C. (2011). A review of data mining applications for quality improvement in manufacturing industry. *Expert Systems with Applications, 38*(10), 13448–13467.

Koo, D., Piratla, K. C., & Matthews, J. (2015). Towards sustainable water supply: Schematic development of big data collection using Internet of Things (IoT). *Procedia Engineering, 118*, 489–497.

Kowalczyk, M., & Buxmann, P. (2015). An ambidextrous perspective on business intelligence and analytics support in decision processes: Insights from a multiple case study. *Decision Support Systems, 80*, 1–13.

Kretschmer, R., Pfouga, A., Rulhoff, S., & Stjepandić, J. (2017). Knowledge-based design for assembly in agile manufacturing by using Data Mining methods. *Advanced Engineering Informatics, 33*, 285–299.

Krumeich, J., Werth, D., & Loos, P. (2016). Prescriptive control of business processes. *Business and Information Systems Engineering, 58*(4), 261–280.

Kubáč, L. (2016). The application of internet of things in logistics. *Transport & Logistics, 16*(39), 9–18.

Kuester, S., & Rauch, A. (2016). A job demands-resources perspective on salespersons' market intelligence activities in new product development. *Journal of Personal Selling & Sales Management, 36*(1), 19–39.

Kumar, A., Shankar, R., Choudhary, A., & Thakur, L. S. (2016). A big data MapReduce framework for fault diagnosis in cloud-based manufacturing. *International Journal of Production Research, 54*(23), 7060–7073.

Kumar, A., Shankar, R., & Thakur, L. (2017). A big data driven sustainable manufacturing framework for condition-based maintenance prediction. *Journal of Computational Science, 27*, 428–439.

Kuo, R. J., Pai, C. M., Lin, R. H., & Chu, H. C. (2015). The integration of association rule mining and artificial immune network for supplier selection and order quantity allocation. *Applied Mathematics and Computation, 250*, 958–972.

Kwak, D. S., & Kim, K. J. (2012). A data mining approach considering missing values for the optimization of semiconductor-manufacturing processes. *Expert Systems with Applications, 39*(3), 2590–2596.

Lade, P., Rumi, G., & Soundar, S. (2017). Manufacturing analytics and industrial internet of things. *IEEE Intelligent Systems, 32*(3), 74–79.

Lamba, K., & Singh, S. P. (2017). Big data in operations and supply chain management: current trends and future perspectives. *Production Planning and Control, 28*, 11–12.

Lanka, S., & Jena, S. K. (2014). *A study on time based association rule mining on spatial-temporal data for intelligent transportation applications.* In Conference on Networks and Soft Computing (ICNSC) (pp. 395–399)

Lau, R. Y. K., Li, C., & Liao, S. S. Y. (2014). Social analytics: Learning fuzzy product ontologies for aspect-oriented sentiment analysis. *Decision Support Systems, 65*, 80–94.

Lee, C. K. H. (2016). A GA-based optimization model for big data analytics supporting anticipatory shipping in Retail 4.0. *International Journal of Production Research, 54*, 1–13.

Lee, J. H., & Chang, M. L. (2010). Stimulating designers' creativity based on a creative evolutionary system and collective intelligence in product design. *International Journal of Industrial Ergonomics, 40*(3), 295–305.

Lee, J., Lapira, E., Bagheri, B., & Kao, H. (2013). Recent advances and trends in predictive manufacturing systems in big data environment. *Manufacturing Letters, 1*(1), 38–41.

Lee, H., Aydin, N., Choi, Y., Lekhavat, S., & Irani, Z. (2017). A decision support system for vessel speed decision in maritime logistics using weather archive big data. *Computers and Operations Research, 98*, 330–342.

Levner, E., Ceder, A., Elalouf, A., Hadas, Y., & Shabtay, D. (2011). *Detection and improvement of deficiencies and failures in public-transportation networks using agent-enhanced distribution data mining.* In IEEE international conference on Industrial Engineering and Engineering Management (IEEM) (pp. 694–698).

Li, H., Parikh, D., He, Q., Qian, B., Li, Z., Fang, D., & Hampapur, A. (2014). Improving rail network velocity: A machine learning approach to predictive maintenance. *Transportation Research Part C: Emerging Technologies*, 17–26.

Li, L., Su, X., Wang, Y., Lin, Y., Li, Z., & Li, Y. (2015). Robust causal dependence mining in big data network and its application to traffic flow predictions. *Transportation Research Part C: Emerging Technologies, 58*, 292–307.

Li, B., Cheng, E., Chong, A. Y., & Bao, H. (2016a). Predicting online e-marketplace sales performances: A big data approach. *Computers and Industrial Engineering, 101*, 565–571.

Li, J., Moghaddam, M., & Nof, S. Y. (2016b). Dynamic storage assignment with product affinity and ABC classification – A case study. *The International Journal of Advanced Manufacturing Technology, 84*(9-12), 2179–2194.

Li, X., Song, J., & Huang, B. (2016c). A scientific workflow management system architecture and its scheduling based on cloud service platform for manufacturing big data analytics. *International Journal of Advanced Manufacturing Technology, 84*(1–4), 119–131.

Lin, J., Xu, X., & Xu, D. M. (2010). Strategic supplier selection: A domain driven data mining methodology. *Information – An International Interdisciplinary Journal, 13*(4), 1449–1465.

Liu, Z., Yumo, W., Cai, L., Cheng, Q., & Zhang, H. (2016). Design and manufacturing model of customized hydrostatic bearing system based on cloud and big data technology. *International Journal of Advanced Manufacturing Technology, 84*(1-4), 261–273.

Lockamy, A., & McCormack, K. (2004). Linking SCOR planning practices to supply chain performance: An exploratory study. *International Journal of Operations & Production Management, 24*(11/12), 1192–1218.

Ma, J., Kwak, M., & Kim, H. M. (2014). Demand trend mining for predictive life cycle design. *Journal of Cleaner Production, 68*, 189–199.

Manyika, J., Chui, M., Brown, B., Bughin, J., Dobbs, R., Roxburgh, C., Byers, H. A., & and. (2011). *Big data: The next frontier for innovation, competition, and productivity* (pp. 1–156). McKinsey Global Institute.

Mariadoss, B. J., Milewicz, C., Lee, S., & Sahaym, A. (2014). Salesperson competitive intelligence and performance: The role of product knowledge and sales force automation usage. *Industrial Marketing Management, 43*(1), 136–145.

Marine-Roig, E., & Clavé, S. A. (2015). Tourism analytics with massive user-generated content: A case study of Barcelona. Journal of Destination Marketing & Management, *4*(3), 162–172.

Markham, S. K., Kowolenko, M., & Michaelis, T. L. (2015). Unstructured text analytics to support new product development decisions. *Research-Technology Management, 58*(2), 30–39.

Mason, R. J., Rahman, M. M., & Maw, T. M. M. (2017). Analysis of the manufacturing signature using data mining. *Precision Engineering, 47*, 292–302.

Mayring, P. (2003). *Qualitative content analysis*. Beltz Verlag.

Mehmood, R., Meriton, R., Graham, G., Hennelly, P., & Kumar, M. (2017). Exploring the influence of big data on city transport operations: A Markovian approach. *International Journal of Operations & Production Management, 37*(1), 75–104.

Min, H. (2010). Artificial intelligence in supply chain management: Theory and applications. *International Journal of Logistics Research and Applications, 13*(1), 13–39.

Miroslav, M., Miloš, M., Velimir, Š., Božo, D., & Đorđe, L. (2014). Semantic technologies on the mission: Preventing corruption in public procurement. *Computers and Industrial Engineering, 65*(5), 878–890.

Mishra, D., Gunasekaran, A., Papadopoulos, T., & Childe, S. J. (2016). Big data and supply chain management: A review and bibliometric analysis. *Annals of Operations Research*, 1–24.

Miyaji, M. (2015). *Data mining for safety transportation by means of using internet survey*. In IEEE international conference on data engineering workshop (pp. 119–123).

Mori, J., Kajikawa, Y., Kashima, H., & Sakata, I. (2012). Machine learning approach for finding business partners and building reciprocal relationships. *Expert Systems with Applications, 39*(12), 10402–10407.

Mourtzis, E. D., Vlachou, E., & Milas, N. (2016). Industrial big data as a result of IoT adoption in manufacturing. *Procedia CIRP, 55*, 290–295.

Moyne, J., Samantaray, J., & Armacost, M. (2017). Big data capabilities applied to semiconductor manufacturing advanced process control. *IEEE Transactions on Semiconductor Manufacturing, 29*(4), 283–291.

Munro, D. L., & Madan, M. S. (2016). Is data mining of manufacturing data beyond first order analysis of value? A case study. *Journal of Decision Systems, 25*(1), 572–577.

Nguyen, T., Li, Z., Spiegler, V., Ieromonachou, P., & Yong, L. (2017). A big data analytics in supply chain management: A state-of-the-art literature review. *Computers and Operations Research*, 1–11.

O'Brien, O., Cheshire, J., & Batty, M. (2014). Mining bicycle sharing data for generating insights into sustainable transport systems. *Journal of Transport Geography, 34*, 262–273.

Olson, D. L. (2015). A review of supply chain data mining publications. *Journal of Supply Chain Management, 9*, 1–13.

Oruezabala, G., & Rico, J. C. (2012). The impact of sustainable public procurement on supplier management – The case of French public hospitals. *Industrial Marketing Management, 41*(4), 573–580.

Ostrowski, D., Rychtyckyj, N., MacNeille, P., & Kim, M. (2016). *Integration of big data using semantic web technologies, Semantic Computing (ICSC)*. In 2016 IEEE Tenth international conference (pp. 382–385).

Packianather, M. S., Davies, A., Harraden, S., Soman, S., & White, J. (2017). Data mining techniques applied to a manufacturing SME. *Procedia CIRP, 62*, 123–128.

Pang, K. W., & Chan, H. L. (2017). Data mining-based algorithm for storage location assignment in a randomised warehouse. *International Journal of Production Research, 55*(14), 4035–4052.

Papadopoulos, T., Gunasekaran, A., Dubey, R., Altay, N., Childe, S., & Fosso-Wamba, S. (2017). The role of big data in explaining disaster resilience in supply chains for sustainability. *Journal of Cleaner Production, 142*(2), 1108–1118.

Park, H., Bellamy, M. A., & Basole, R. C. (2016). Visual analytics for supply network management: System design and evaluation. *Decision Support Systems, 91*, 89–102.

Peters, H., & Link, N. (2010). Cause and effect analysis of quality deficiencies at steel production using automatic data mining technologies. *IFAC Proceedings Volumes, 43*(9), 56–61.

Petri, M., Pratelli, A., & Fusco, G. (2016). Data mining and big freight transport database analysis and forecasting capabilities. *Transactions on Maritime Science-TOMS, 5*(2), 99–110.

Prasad, S., Zakaria, R., & Altay, N. (2016). Big data in humanitarian supply chain networks: A resource dependence perspective. *Annals of Operations Research.* https://doi.org/10.1007/s10479-016-2280-7

Ralha, C. G., Silva, S., & Vinicius, C. (2012). A multi-agent data mining system for cartel detection in Brazilian government procurement. *Expert Systems with Application, 39*(14), 11642–11656.

Reuter, C., Brambring, F., Weirich, J., & Kleines, A. (2016). Improving data consistency in production control by adaptation of data mining algorithms. *Procedia CIRP, 56*, 545–550.

Richey, R. G., Jr., Morgan, T. R., Morgan, T. R., Lindsey-Hall, K., Lindsey-Hall, K., & Adams, F. G. (2016). A global exploration of big data in the supply chain. *International Journal of Physical Distribution and Logistics Management, 46*(8), 710–739.

Robinson, D. C., Sanders, D. A., & Mazharsolook, E. (2015). Ambient intelligence for optimal manufacturing and energy efficiency. *Assembly Automation, 35*(3), 234–248.

Ronowicz, J., Thommes, M., Kleinebudde, P., & Krysiński, J. (2015). A data mining approach to optimize pellets manufacturing process based on a decision tree algorithm. *European Journal of Pharmaceutical Sciences, 73*, 44–48.

Salehan, M., & Kim, D. J. (2016). Predicting the performance of online consumer reviews: A sentiment mining approach to big data analytics. *Decision Support Systems, 81*, 30–40.

Sanders, N. R. (2016). How to use big data to drive your supply chain. *California Management Review, 58*(3), 26–48.

Sangari, M. S., & Razmi, J. (2015). Business intelligence competence, agile capabilities, and agile performance in supply chain: An empirical study. *The International Journal of Logistics Management, 26*(2), 356–380.

Sann, A., Krimmling, J., Baier, D., & Ni, M. (2013). Lead user intelligence for complex product development: the case of industrial IT–security solutions. *International Journal of Technology Intelligence and Planning, 9*(3), 232–249.

Schmidt, B., Gandhi, K., Wang, L., & Galar, D. (2017). Context preparation for predictive analytics – A case from manufacturing industry. *Journal of Quality in Maintenance Engineering, 23*(3), 341–354.

Schoenherr, T., & Speier-Pero, C. (2015). Data science, predictive analytics, and big data in supply chain management: current state and future potential. *Journal of Business Logistics, 36*(1), 120–132.

Schoenherr, T., & Swink, M. (2015). The roles of supply chain intelligence and adaptability in new product launch success. *Decision Sciences, 46*(5), 901–936.

Seuring, S., & Müller, M. (2008). From a literature review to a conceptual framework for sustainable supply chain management. *Journal of Cleaner Production, 16*, 1699–1710.

Shafiq, S. I., Sanin, C., Szczerbicki, E., & Carlos, T. (2017). Towards an experience based collective computational intelligence for manufacturing. *Future Generation Computer Systems, 66*, 89–99.

Shan, Z., & Zhu, Q. (2015). Camera location for real-time traffic state estimation in urban road network using big GPS data. *Neurocomputing, 169*, 134–143.

Shanmugasundaram, P., & Paramasivam, I. (2016). Big data analytics bring new insights and higher business value – An experiment carried out to divulge sales forecasting solutions. *International Journal of Advanced Intelligence Paradigms, 8*(2), 207–218.

Shi, Q., & Abdel-Aty, M. (2015). Big data applications in real-time traffic operation and safety monitoring and improvement on urban expressways. *Transportation Research Part C: Emerging Technologies, 58*, 380–394.

Shin, S. J., Woo, J., & Sudarsan, R. (2014). Predictive analytics model for power consumption in manufacturing. *Procedia CIRP, 15*, 153–158.

Shukla, N., & Kiridena, S. (2016). A fuzzy rough sets-based multi-agent analytics framework for dynamic supply chain configuration. *International Journal of Production Research, 54*(23), 6984–6996.

Sivamani, S., Kwak, K., & Cho, Y. (2014). A study on intelligent user-centric logistics service model using ontology. *Journal of Applied Mathematics*, 1–10.

Soban, D., Thornhill, D., Salunkhe, S., & Long, A. (2016). Visual analytics as an enabler for manufacturing process decision-making. *Procedia CIRP, 56*, 209–214.

Sodhi, M. S., & Tang, C. S. (2011). Determining supply requirement in the sales-and operations-planning (S&OP) process under demand uncertainty: A stochastic formulation and a spreadsheet implementation. *Journal of the Operational Research Society, 62*(3), 526–536.

Soroka, A., Liu, Y., Han, L., Haleem, M., & S. (2017). Big data driven customer insights for SMEs in redistributed manufacturing. *Procedia CIRP, 63*, 692–697.

Souza, G. C. (2014). Supply chain analytics. *Business Horizons, 57*(5), 595–605.

Srinivasan, R., & Swink, M. (2017, 2017). An investigation of visibility and flexibility as complements to supply chain analytics: An organizational information processing theory perspective. *Production and Operations Management*. https://doi.org/10.1111/poms.12746

St-Aubin, P., Saunier, N., & Miranda-Moreno, L. (2015). Large-scale automated proactive road safety analysis using video data. *Transportation Research Part C: Emerging Technologies, 58*, 363–379.

Stefanovic, N. (2015). Collaborative predictive business intelligence model for spare parts inventory replenishment. *Computer Science and Information Systems, 12*(3), 911–930.

Stewart, G. (1997). Supply-chain operations reference model (SCOR): The first cross-industry framework for integrated supply-chain management. *Logistics Information Management, 10*(2), 62–67.

Tachizawa, E. M., Alvarez-Gil, M. J., & Montes-Sancho, M. J. (2015). How "smart cities" will change supply chain management. *Supply Chain Management: An International Journal, 20*(3), 237–248.

Tan, K. H., Zhan, Y. Z., Ji, G., Ye, F., & Chang, C. (2015). Harvesting big data to enhance supply chain innovation capabilities: An analytic infrastructure based on deduction graph. *International Journal of Production Economics, 165*(2015), 223–233.

Tanev, S., Liotta, G., & Kleismantas, A. (2015). A business intelligence approach using web search tools and online data reduction techniques to examine the value of product-enabled services. *Expert Systems with Applications, 42*(21), 7582–7600. https://doi.org/10.1016/j.eswa.2015.06.006

Theeranuphattana, A., & Tang, J. C. S. (2008). A conceptual model of performance measurement for supply chains alternative considerations. *Journal of Manufacturing Technology Management, 19*(1), 125–148.

Thiruverahan, N., & Subramanian, N. (2015). *Data mining and machine learning based approach to inventory prediction-a case study*. In International conference on Control, Automation and Artificial Intelligence (CAAI) (pp. 195–202).

Thotappa, C., & Ravindranath, K. (2010). Data mining aided proficient approach for optimal inventory control in supply chain management. *World Congress on Engineering (WCE), 1*, 341–345.

Toole, J. L., Colak, S., Sturt, B., Alexander, L. P., Evsukoff, A., & González, M. C. (2015). The path most traveled: Travel demand estimation using big data resources. *Transportation Research Part C: Emerging Technologies, 58*, 162–177.

Trkman, P., McCormack, K., de Oliveira, M. P. V., & Ladera, M. B. (2010). The impact of business analytics on supply chain performance. *Decision Support Systems, 49*, 318–327.

Tsai, C. Y., & Huang, S. H. (2015). A data mining approach to optimise shelf space allocation in consideration of customer purchase and moving behaviours. *International Journal of Production Research, 53*(3), 850–866.

Tsao, Y.-C. (2017). Managing default risk under trade credit: Who should implement big-data analytics in supply chains. *Transportation Research Part E: Logistics and Transportation Review, 106*, 276–293.

Tsuda, T., Inoue, S., Kayahara, A., Imai, S., Tanaka, T., Sato, N., & Yasuda, S. (2015). Advanced semiconductor manufacturing using big data. *IEEE Transactions on Semiconductor Manufacturing, 28*(3), 229–235.

Tu, W., Li, Q., Fang, Z., Shaw, S., Zhou, B., & Chang, X. (2016). Optimizing the locations of electric taxi charging stations: A spatial–temporal demand coverage approach. *Transportation Research Part C: Emerging Technologies, 65*, 172–189.

Ulrike, F., Dannecker, L., Siksnys, L., Rosenthal, F., Boehm, M., & Wolfgang, L. (2013). Towards integrated data analytics: Time series forecasting in DBMS. *Datenbank-Spektrum, 13*(1), 45–53.

Ünay, F. G., & Zehir, C. (2012). Innovation intelligence and entrepreneurship in the fashion industry. *Procedia-Social and Behavioral Sciences, 41*, 315–321.

ur Rehman, M. H., Chang, V., Batool, A., & Wah, T. Y. (2016). Big data reduction framework for value creation in sustainable enterprises. *International Journal of Information Management, 36*(6), 917–928.

van der Spoel, S., Amrit, C., & van Hillegersberg, J. (2017). Predictive analytics for truck arrival time estimation: A field study at a European distribution centre. *International Journal of Production Research, 55*(17), 5062–5078.

Veugelers, M., Bury, J., & Viaene, S. (2010). Linking technology intelligence to open innovation. *Technological Forecasting and Social Change, 77*(2), 335–343.

Walker, H., & Brammer, S. (2012). The relationship between sustainable procurement and e-procurement in the public sector. *International Journal of Production Economics, 140*(1), 256–268.

Walker, G., & Strathie, A. (2016). Big data and ergonomics methods: A new paradigm for tackling strategic transport safety risks. *Applied Ergonomics, 53*, 298–311.

Wallander, J., & Makitalo, M. (2012). Data mining in rail transport delay chain analysis. *International Journal of Shipping and Transport Logistics, 4*(3), 269–285.

Waller, M. A., & Fawcett, S. E. (2013). Data science, predictive analytics, and big data: A revolution that will transform supply chain design and management. *Journal of Business Logistics, 34*(2), 77–84.

Wamba, S. F., Akter, S., Edwards, A., Chopin, G., & Gnanzou, D. (2015). How 'big data' can make big impact: Findings from a systematic review and a longitudinal case study. *International Journal of Production Economics, 165*, 234–246.

Wang, X. K., & Yang, L. (2016). Visual data mining in transportation using multiresolution data aggregation. *Fuzzy System and Data Mining, 281*, 195–199.

Wang, J., & Zhang, J. (2016). Big data analytics for forecasting cycle time in semiconductor wafer fabrication system. *International Journal of Production Research, 54*(23), 7231–7244.

Wang, Y., Shao, Y., Matovic, M. D., & Whalen, J. K. (2016a). Recycling combustion ash for sustainable cement production: A critical review with data-mining and time-series predictive models. *Construction and Building Materials, 123*, 673–689.

Wang, C., Li, X., Zhou, X., Wang, A., & Nedjah, N. (2016b). Soft computing in big data intelligent transportation systems. *Applied Soft Computing, 38*, 1099–1108.

Wang, G., Gunasekaran, A., Ngai, E. W. T., & Papadopoulos, T. (2016c). Big data analytics in logistics and supply chain management: Certain investigations for research and applications. *International Journal of Production Economics, 176*, 98–110.

Wang, S., Wan, J., Zhang, D., Li, D., & Zhang, C. (2016d). Towards smart factory for industry 4.0: A self-organized multi-agent system with big data based feedback and coordination. *Computer Networks, 101*, 158–168.

Wang, J., Zhang, L., Duan, L., & Gao, R. X. (2017). A new paradigm of cloud-based predictive maintenance for intelligent manufacturing. *Journal of Intelligent Manufacturing, 28*(5), 1125–1137.

Westerski, A., Kanagasabai, R., Wong, J., & Chang, H. (2015). Prediction of enterprise purchases using markov models in procurement analytics applications. *Procedia Computer Science, 60,* 1357–1366.

Wiener, L. O., & Julia, O. (2010). Building process understanding for vaccine manufacturing using data mining. *Quality Engineering, 22*(3), 1–30.

Williams, B. D., Roh, J., Tokar, T., & Swink, M. (2013). Leveraging supply chain visibility for responsiveness: The moderating role of internal integration. *Journal of Operations Management, 31*(7–8), 543–554.

Wu, P., Huang, Y. F., Cao, Q. Q., & Xiong, F. (2013). Research on mining of E-procurement model parameters based on decision tree. *Advanced Design and Manufacturing, 397,* 2655–2661.

Wu, K., Liao, C., Tseng, M., Lim, M. K., Hu, J., & Tan, K. (2017a). Toward sustainability: Using big data to explore the decisive attributes of supply chain risks and uncertainties. *Journal of Cleaner Production, 142,* 663–676.

Wu, P.-J., Chen, M., & Tsau, C. (2017b). The data-driven analytics for investigating cargo loss in logistics systems. *International Journal of Physical Distribution and Logistics Management, 47*(1), 68–83.

Xiao, S., Wei, C.-P., & Dong, M. (2015). Crowd intelligence: Analyzing online product reviews for preference measurement. *Information and Management, 53*(2), 169–182.

Xie, X. L. (2016). Research on data mining model of intelligent transportation based on granular computing. *International Journal of Security and Its Applications, 10*(7), 281–286.

Xu, J., & Güting, R. H. (2013). A generic data model for moving objects. *GeoInformatica, 17*(1), 125–172.

Xu, Z., Frankwick, G. L., & Ramirez, E. (2016). Effects of big data analytics and traditional marketing analytics on new product success: A knowledge fusion perspective. *Journal of Business Research, 69*(5), 1562–1566.

Yeniyurt, S., Henke, J. W., & Cavusgil, E. (2013). Integrating global and local procurement for superior supplier working relations. *International Business Review, 22*(2), 351–362.

Zaki, M., Theodoulidis, B., Shapira, P., Neely, A., & Surekli, E. (2017). The role of big data to facilitate redistributed manufacturing using a co-creation lens: Patterns from consumer goods. *Procedia CIRP, 63,* 680–685.

Zangenehpour, S., Miranda-Moreno, L. F., & Saunier, N. (2015). Automated classification based on video data at intersections with heavy pedestrian and bicycle traffic: Methodology and application. *Transportation Research Part C: Emerging Technologies, 56,* 161–176.

Zhan, Y., Tan, K. H., Ji, G., Chung, L., & Tseng, M. (2017). A big data framework for facilitating product innovation processes. *Business Process Management Journal, 23*(3), 518–536.

Zhang, Y., Zhang, G., Du, W., Wang, J., Ali, E., & Sun, S. (2015). An optimization method for shop floor material handling based on real-time and multi-source manufacturing data. *International Journal of Production Economics, 165,* 282–292.

Zhang, Y., Ren, S., Liu, Y., & Si, S. (2017). A big data analytics architecture for cleaner manufacturing and maintenance processes of complex products. *Journal of Cleaner Production, 142,* 626–641.

Zhao, X., & Rosen, D. (2017). A data mining approach in real-time measurement for polymer additive manufacturing process with exposure controlled projection lithography. *Journal of Manufacturing Systems, 43*(2), 271–286.

Zhao, R., Yiyun, L., Zhang, N., & Tao, H. (2017). An optimization model for green supply chain management by using a big data analytic approach. *Journal of Cleaner Production, 142,* 1085–1097.

Zhong, R. Y., Huang, G. Q., & Dai, Q. (2013). Mining standard operation times for real-time advanced production planning and scheduling from RFID-enabled shop floor data. *IFAC Proceedings Volumes, 46*(9), 1950–1955.

Zhong, R. Y., Huang, G. Q., Lan, S. L., Dai, Q. Y., Xu, C., & Zhang, T. (2015a). A big data approach for logistics trajectory discovery from RFID-enabled production data. *International Journal of Production Economics, 165,* 260–272.

Zhong, R. Y., Huang, G. Q., Lan, S., Dai, Q. Y., Zhang, T., & Xu, C. (2015b). A two-level advanced production planning and scheduling model for RFID-enabled ubiquitous manufacturing. *Advanced Engineering Informatics, 29*(4), 799–812.

Zhong, R. Y., Stephen, T. N., George, Q. H., & Shulin, L. (2016a). Big data for supply chain management in the service and manufacturing sectors: Challenges, opportunities, and future perspectives. *Computers and Industrial Engineering, 101*, 572–591.

Zhong, R. Y., Lan, S., Xu, C., Dai, Q., & Huang, G. Q. (2016b). Visualization of RFID-enabled shop floor logistics big data in cloud manufacturing. *The International Journal of Advanced Manufacturing Technology, 84*(1–4), 5–16.

Zhong, R. Y., Xu, C., Chen, C., & Huang, G. Q. (2017). Big data analytics for physical internet-based intelligent manufacturing shop floors. *International Journal of Production Research, 55*(9), 2610–2621.

Zhou, G. (2016). Research on supplier performance evaluation system based on data mining with triangular fuzzy information. *Journal of Intelligent and Fuzzy Systems, 31*(3), 2035–2042.

Zhou, M., Wang, D. G., Li, Q. Q., Yue, Y., Tu, W., & Cao, R. (2017). Impacts of weather on public transport ridership: Results from mining data from different sources. *Transportation Research Part C-Emerging Technologies, 75*, 17–29.

Zhu, H. P., Xu, Y., Liu, Q., & Rao, Y. Q. (2014). Cloud service platform for big data of manufacturing. *Applied Mechanics and Materials, 456*, 178–183.

Unveiling the Role of Evolutionary Technologies for Building Circular Economy-Based Sustainable Manufacturing Supply Chain

Rashmi Prava Das, Kamalakanta Muduli, Sonia Singh, Bikash Chandra Behera, and Adimuthu Ramasamy

1 Introduction

The fourth industrial revolution (Industry 4.0) has emerged as a significant facilitator for manufacturing companies in a world marked by globalization, increasing consumer demand, increased competition, changing consumer behaviour, limited resources, environmental pollution caused by the disposal of end products and rising production costs. Manufacturers' activities are being transformed as a result of the advancement of information communication technology (ICT) and data storage, and the third industrial evolution (Nascimento et al., 2019) has arrived at the right time to assist in the development and adoption of the circular economy concept (CEC). Rapid delivery, automated and high-quality products, as well as the desire of consumers for more customized products are driving businesses towards the fourth industrial revolution (Industry 4.0) (Zheng et al., 2021). Manufacturing will be transformed in the future as a result of networking within an Internet of Things,

R. P. Das · B. C. Behera
Department of Computer Science, CV Raman Global University, Bhubaneswar, Odisha, India

K. Muduli (✉)
Department of Mechanical Engineering, Papua New Guinea University of Technology, Lae, Morobe Province, Papua New Guinea

S. Singh
Director Toss Global Management, Dubai, UAE

A. Ramasamy
Department of Business Studies, Papua New Guinea University of Technology, Lae, Morobe Province, Papua New Guinea

© The Author(s), under exclusive license to Springer Nature Switzerland AG 2023
S. S. Kamble et al. (eds.), *Digital Transformation and Industry 4.0 for Sustainable Supply Chain Performance*, EAI/Springer Innovations in Communication and Computing, https://doi.org/10.1007/978-3-031-19711-6_2

services, data and people. Manufacturing companies must adapt to the digital transformation of changes in their manufacturing processes if they are to remain competitive in the marketplace.

There are a good number of studies (Ivanov et al., 2016; Kang et al., 2016; Lom et al., 2016; Wan et al., 2016a, b) on the topic of Industry 4.0 which have proposed several benefits from the implementation of Industry 4.0. Some of the common benefits include optimized production, increased productivity, customization of products and delivering it to a value-added end user, improved control of data operations, improved communication and collaboration in supply chain management, reduced cost and increased profit and subsequent shareholder value, creating innovation and opportunities.

The new circular economic concept has been developed in response to the need for improvement as well as the opportunity provided by Industry 4.0. The CEC integrates Industry 4.0 techniques with circular economic practices to provide a sustainable service in the supply chain management to fulfil consumer requirements. To meet the current demands of society, Parajuly and Wenzel (2017) and Sousa-Zomer and Miguel (2018) advocated the need and opportunity for a new path of economic development, which included dealing with waste generated by society in a new circular business model that reduces, recycles and repurposes such waste with the goal of transforming it into higher value-added products to meet the current demands of society. Mastos et al. (2021) provided in their study that the solutions that incorporate Industry 4.0 technologies are key factors for a successful development of circular economy concept.

The application of Industry 4.00 techniques in the context of the circular economy concept has gained widespread acceptance among manufacturers and governments in both developed and developing countries. However, in some developing countries, most manufacturing companies are still working on the traditional linear economy model (Kumar et al. 2021a, b). According to interviews, Papua New Guinea (PNG) is one of the countries with a small number of manufacturing companies that use the recycle process to repurpose their products.

At the stage of digitalization with fourth industrial evolution enabling and transforming manufacturing industries to adopt CEC, this study focuses on addressing on how the Industry 4.0 techniques enable the manufacturing industries to create a promising CEC and how effective manufacturing industries adopting to CEC as to avoid being left behind, especially in developing countries like Papua New Guinea (PNG). There have been a number of previous studies that have addressed the application of Industry 4.0 techniques in CEC for manufacturing industries, but there has not been much emphasis placed on the urgency for manufacturing industries in developing countries such as PNG to equally adopt the CEC, despite the fact that developing countries are members of the global community. Therefore, the following two research questions were formed to show the manufacturing industries, especially in developing countries the usefulness of applying Industry 4.0 techniques in their factories to create a more promising CEC.

RQ1: What role can Industry 4.0 technologies play in transforming the activities of manufacturing industries in order to create a more promising circular economy concept?

RQ2: How effective are manufacturing industries, particularly in developing countries such as Papua New Guinea, in adopting CEC through the use of Industry 4.0 techniques?

The following is the order in which the paper is organized in order to respond to the questions above: Section 2 describes the literature review on the past studies, Sect. 3 provides the qualitative research method, Sect. 4 shows the results and discussions, followed by Sect. 5 with limitations and recommendations of the study and Sect. 6 with conclusions and, finally, references.

2 Literature Review on Industry 4.0 – Manufacturing and Circular Economy Concept

2.1 Industry 4.0 (Fourth Industrial Revolution)

The study by Deloitte showed that there are four main characteristics related to Industry 4.0 that point out the direction manufacturers should take to operate in a very competitive business environment. These characteristics include vertical networking, horizontal integration, through-engineering and acceleration through exponential technologies (Deloitte).

Vertical Networking of Smart Production Systems
This is about vertical connectivity of smart production technologies, such as smart factories and smart goods, as well as smart logistics, manufacturing, marketing and smart services, with a strong needs-oriented, personalized and customer-specific manufacturing operation.

The Horizontal Integration
The **Horizontal Integration** is new generation of global value creation networks, which includes the integration of business partners and customers, as well as new business and collaboration models that span nations and models.

Through-Engineering
Through Modern Engineering Practices, industries can focus on value addition across the whole supply chain rather than only in the manufacturing process. In simple words the focus should be shifted from only the final product to the complete product life cycle the product.

Acceleration Through Exponential Technologies
Acceleration by exponential technologies can enable mass marketing applications as their cost and size have decreased, such as sensor technology, and their processing power has increased dramatically.

Industry 4.0 overview I Deloitte Insights: (Accessed on 6th Dec 2021)

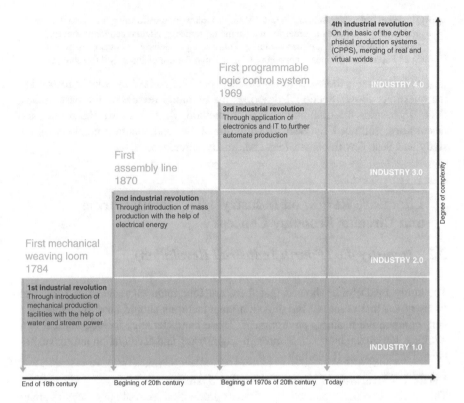

Fig. 1 showing the development and timeline of the evolution of manufacturing and industrial sector. (Source: Deloitte)

In 2011, a group of professionals from diverse sectors of specialism in Germany launched the creation of Industry 4.0. The goal was to boost the competitiveness of Germany's industrial industry. Industry 4.0 grew out of Industry 3.0 and the change of information and communication technologies (ICT). Figure 1 depicts the history and phases of development of four industries. The following are concise definitions of the other three industrial revolutions:

Industry 1.0
In the eighteenth century (1760–1840), the first industrial revolution was introduced in Britain, allowing machines to produce with the assistance of water and steam.

Industry 2.0
The second industrial revolution was established in the nineteenth century and lasted until the early twentieth century (1870–1914). Electricity made it possible to make a lot of things quickly during this time.

Industry 3.0
It was during the twentieth century (1950–1970) that the third industrial revolution began, with the application of electronic and information technology to further

automate production. This occurred at the same time as another significant revolution, the digital revolution, which was brought about by rapid advancements in computer and information communication technology (ICT).

2.1.1 Industry 4.0 Design Principles

Manufacturers are encouraged to investigate the possibility of transitioning to Industry 4.0 technologies through the application of six Industry 4.0 design principles developed on the basis of value chain creation. In their study, Zheng et al. (2021) also made a note of the design principles that were observed. The findings of this study demonstrate that Industry 4.0 is an enabler for manufacturers and that it guides them through the adoption process.

Interoperability

To conform with the interoperability principle, objects, machines and persons must be able to connect with one another via IoT and the IoP. As a consequence, a smart factory is created.

Virtualization

According to virtualization requirements, CPS must be able to mimic and produce a virtual duplicate of the real environment. Furthermore, CPS must be capable of monitoring items in the surrounding environment.

Decentralization

The decentralization principle provides the ability for CPS to operate independently of the rest of the organization. This allows for greater customization of products as well as problem-solving opportunities. Additionally, it creates a flexible environment for manufacturing. Failure and conflicting objectives result in the issue being delegated to a higher level of authority (Joshi et al., 2022; Muduli et al., 2022).

Capability to Operate in Real Time

Real-time capability is intended to allow smart factories to collect real-time data, store and analyse it and make decisions in response to new discoveries. Data collected in real time is used for market research, process improvement and maintenance in internal business processes, such as when a machine fails on a production line. Smart objects must be able to detect and re-delegate tasks to other operating machines in the event of a malfunction. The result is increased flexibility and optimization of the manufacturing process.

Providing Excellent Customer Service

Productions must be oriented towards the customer in order to be considered service-oriented. In order to build goods that are personalized to the requirements of individual consumers, persons and intelligent devices must be able to link effectively through to the Internet of Services.

Modularity
A smart factory's ability to adapt to new markets in a dynamic market is facilitated by the modularity principle. Smart factories (companies) typically take a week to study the market and make changes to their production in response to the findings of their research. The ability to adapt quickly and smoothly to seasonal changes and market trends is also essential for smart factories to succeed.

2.1.2   Techniques for Enabling Industry 4.0

Industry 4.0 techniques are widely accepted and implemented in manufacturing industries; however, the development of enabling technologies and research is still ongoing; as a result, various publications have stated that there is no agreed-upon number of Industry 4.0 techniques (Zheng et al., 2021). The following common Industry 4.0 techniques were selected from the various literatures cited in the studies of common Industry 4.0 techniques (Zheng et al., 2021; Nascimento et al., 2019).

Cyber-Physical Systems (CPS)

In computing, a CPS is a grouping of transformational technologies which are aimed at integrating computation with physical assets. The primary function of computers and networks in manufacturing is to monitor the physical systems (processes) involved in the manufacturing process (Lee et al., 2015; Monostori et al., 2016). CPS is comprised of three phases that must be completed in order for it to be effective. Identification, the integration of sensors and actuators and the development of sensors and actuators are some of the tasks involved. RFID (Radio Frequency Identification) is an example of how unique identification is important in the manufacturing industry. A process in which the movement of the machine can be controlled and monitored in order to detect changes in the process, but which does not allow the sensors and actuators to communicate with one another, is known as sensor and actuator integration. Because of the advancement of sensors and actuators, CPS is able to store and analyse data for the purpose of improving and exchanging information. CPS also creates smart networks of machines, buildings, information and communication technology systems, smart goods and people that span the full value chain and product life cycle.

The Internet of Things (IoT)

The Internet of Things (IoT) is a network of physical objects such as sensors, mobile phones, machines and automobiles that communicate with one another and with humans in order to allow interaction and cooperation in order to come up with a solution to a problem (Oztemel and Gursev 2020; Trappey et al. 2016).

Analytics and Data on the Internet of Things (IoT)

The Internet of Big Data and Analytics (IoBDA) enables the collecting and analysis of huge volumes of accessible data by utilizing a range of strategies to filter, capture and publish insights when enormous amounts of data are processed (Wamba et al. 2015; Vera-Baquero et al. 2014).

The phrase 'smart factory' refers to a factory that is software in its operations.

A technique known as the calm system is used in smart factories, which is capable of dealing with both the physical and virtual worlds. A calm system is in operation behind the scenes, and it is cognizant of its surroundings and the objects in its immediate vicinity. The authors of the study describe a smart factory as 'a factory where CPS interact across the Internet of Things and aid humans and equipment in the execution of their jobs'.

Internet of Services

The Internet of Services connects the customer to the manufacturers through various devices like smart phones and laptops through the Internet. This enables manufacturers to have information on customer demand and behaviour for improvement and to plan for production.

Cloud Technology

Cloud technology is an online system that provides for storing data and programmes for all applications (Zhang et al., 2014; Dong et al., 2014). This service is safe and does not require additional installations.

Blockchain

Blockchain is a distributed and tamper-resistant digital record of transactions that includes timestamps of blocks kept by each participating node (Ghobakhloo 2018; Sikorski et al., 2017; Swain et al., 2021).

Additive Manufacturing (AM)

Additive manufacturing is a technology that involves layering materials to create products using 3D model data, allowing for more design possibilities and mass customization (Esmaeilian et al., 2016). 3D printing technologies provide a number of advantages, including the ability to create things on demand at low prices, the ability to shift upstream or downstream quickly to modify the degree of vertical integration and the ability to make business models flexible and adaptable. AM is also known as 3D printing since it involves the creation of a 3D CAD model.

Artificial Intelligence (AI): Robots

AI is a system that is made to think like human beings and reason out according to six main design principles (Russell and Norvig 2016).

2.1.3 Industry 4.0 Characteristics

The studies (Deloitte) showed that there are four main characteristics related to Industry 4.0 that point out the direction manufacturers should take to operate in a very competitive business environment. These characteristics include vertical networking, horizontal integration, through-engineering and acceleration through exponential technologies (Deloitte).

Vertical Networking of Smart Production Systems

This is about vertical connectivity of smart production technologies, such as smart factories and smart goods, as well as smart logistics, manufacturing, marketing and smart services, with a strong needs-oriented, personalized and customer-specific manufacturing operation.

The Horizontal Integration

Horizontal integration is new generation of global value creation networks, which includes the integration of business partners and customers, as well as new business and collaboration models that span nations and models.

Through Engineering Across the Whole Value Chain

Through value chain re-engineering includes not only the manufacturing process but also the final product – that is, the complete product life cycle of both the product and the customers.

Acceleration Through Exponential Technologies

Acceleration by exponential technologies can enable mass marketing applications as their cost and size have decreased, such as sensor technology, and their processing power has increased dramatically.

2.1.4 The Industry 4.0 Environment

With the development of Industry 4.0 based on the design principles, reflect the above characteristics in the environment of Internet of Things, services, big data and people (Fig. 2).

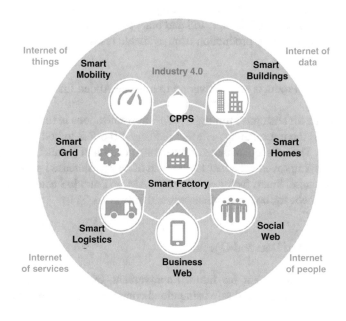

Fig. 2 Industry 4.0 environment

2.1.5 Benefits from the Application of Industry 4.0 Techniques

A good number of studies (Ivanov et al., 2016; Kang et al., 2016 Lom et al., 2016 Wan et al., 2016a; b, Thoben et al., 2017; Li et al., 2017) on the topic of Industry 4.0 have proposed several benefits on the implementation of Industry 4.0. Some common benefits include optimized production, increased productivity, customization of products and delivering to a value-added end user, improved control of data operations, improved communication and collaboration in supply chain management, reduced cost and increased profit and subsequent shareholder value, creating innovation and opportunities.

Optimized Production

Industry 4.0 provides a significant benefit to enterprises in the form of production optimization. Smart factories with thousands of gadgets that can self-optimize output will result in almost no production downtime. According to a PWC research, the ability to use production continuously and consistently will benefit the organization.

Increased Productivity

Increase in demand with customer-oriented products requires smart factories with CPS, AI, AM with 3D models, IoT, IoS and other techniques that communicate and collaborate in maximizing production from available resources.

Customization of Products and Delivering to a Value-Added End User

Increasing value for customers through product customizations in fulfilling customer specifications is the important benefit that Industry 4.0 offers to both manufacturers and customers. The subsequent benefits are strong position of products in the market, increased sales, improved customer loyalty and of course increased profit that helps sustain the business. Smart factories with CPS, AM, IoT and IoS lead in enabling to achieve this benefit for both manufacturers and customers.

Improved Control of Data and Operations

Historical data is important for future improvement, innovation and operations. Therefore, data is being stored safe using cloud computing and Internet of Big data.

Improved Communication and Collaboration in Supply Chain Management

The product is as good as the supply chain management is because without the product reaching the customers for consumption, all costs and effort involved in production are of no value. Transporters, suppliers, manufacturers, distributors, retailers and consumers use cloud computing, IoS and IoT to share information, communicate and build improved relationships and delivery time.

Reduced Costs

Industry 4.0 techniques enable manufacturers to optimize production at reduced costs like labour hours, electricity and even in the supply chain for distribution and sales. IoT and IoS reduced marketing and sales costs and online services help market the product.

Increased Profit and Shareholder Value

Optimization in production, product customization to fulfil customer specifications (customer-oriented products) increased production from smart factories, efficiency in production, reduction in cost of production sharing information and improved network and communication will surely increase sales and subsequently increase profit that contributes to increased share value of owners. Industry 4.0 applications that enable profit maximization are smart factories with CPS, AM, AI, IoT and IoS with all other techniques for both sustainability of business and increased shareholder value.

Creating Innovations and Opportunities

In a competitive business environment, innovation and opportunities created by Industry 4.0 techniques are of huge benefit to manufacturers, researchers and education and potential customers. Manufacturing being creative in product designs and specifications by using smart factories with CPS. It is an opportunity for researchers and education to continue to research for more innovations in business, especially in the manufacturing sector. Consumers enjoy the benefits of innovation and opportunities in terms of increased production and customized products, even accessing to products and services through IoT and IoS from smart phones.

2.1.6   Industry 4.0 Challenges

The challenges in the adoption and application of Industry 4.0 are security, capital investment and training, employment and privacy.

Information Security Risk

The online integration of network, sharing information and data, poses IT risks to the implementation of Industry 4.0. It is likely that there will security breaches and leakage of data. Cyber theft is possible as well. This presents a risk to manufacturers in terms of losing important information, even reputation and will cost the company. This area requires more research to close this known gap.

Capital Costs and Training

Certainly, the transformation by Industry 4.0 in the manufacturing sector requires a huge capital investment for new business model to position in the competitive business environment. Some manufacturers understand the past, present and the future of business that is driven by digital transformation which Industry 4.0 stands as the current enabler for manufacturing industry. However, others fear of investing big money as leaders and key staff in the organization have not much knowledge of the Industry 4.0 techniques and their operations. This leads to investing in training and development on the use and maintenance of Industry 4.0, which is another challenge. This area requires more research, awareness on the benefits of Industry 4.0 that outweigh the capital and training costs in the long run and the manufacturing industry with economy driven by digital transformation.

Employment

It is early to predict but according to the review studies, some countries like China, Germany and the United States are focusing on adopting Industry 4.0 in their manufacturing industries and the business. Industry 4.0 may push some people out of work and require new set of knowledge and skills from staff and management. More research and investment in AI and robotic development in the future may change the current trend of more staff engagement and less use of Industry 4.0 techniques, especially robots. Unemployment will increase when robots and machines do most of the tasks that people are currently doing.

Privacy

Privacy of personal data and information is exposed to others in sharing of information and data through interconnection and network provided by Industry 4.0. Manufacturers may collect and analyse data, but it is a threat to the consumers. Also, some companies who have not been sharing information and data is a challenge for them.

2.2 Manufacturing and the Current Position in the Fourth Industrial Revolution (Industry 4.0)

Manufacturing is the process of creating items with the help of people, machinery, equipment, chemicals and biological processes (formulation). The phrase may be used to a wide variety of human activities, from simple crafts to high-tech, but it is most commonly associated with industrial design, which involves the transformation of raw materials from the primary industries sector into completed commodities on a big scale. The steps that raw materials are changed into completed things are referred to as manufacturing engineering or manufacturing process. To obtain the desired result, the manufacturing process begins with the product design and material specifications. All intermediate processes necessary in the manufacture and integration of product components are included in modern manufacturing.

2.2.1 The Purpose of Manufacturing Industries and Current Position

Manufacturing industries mainly exist to produce goods to meet the ever-growing demand of consumers. Also, as a business, profit is another motive. Increase in competition, change in consumer behaviour, environmental pollution through disposal of waste products and industry evolutions through advancement from information communication technology and scarcity of raw materials and other resources demand manufacturing business to seek an alternative to adopt to continue production and remain in business. The modern manufacturing industries using Industry 3.0 and 4.0 is way forward for the industry. However, the process of adopting especially Industry 4.0 is slow due to various challenges.

2.2.2 Challenges of Manufacturing Industries in Adopting Industry 4.0

Six out of ten manufacturers say that Industry 4.0 implementation difficulties are so high that their Industry 4.0 projects have made only little progress in the previous year. The following are some of the most often mentioned challenges:

- Concerns about data ownership when selecting third-party contractors to host and operate firm data.
- There is a scarcity of in-house talent to assist the development and implementation of Industry 4.0 initiatives. Inability to implement a major digitization strategy due to a lack of bravery.
- The fear of huge capital investment in Industry 4.0 initiatives.

The Industry 4.0 offers solutions to manufacturing companies to adopt, but the barriers restrict and delay the adoption for others. However, it remains a business decision and a choice for any manufacturing entity that examines the trend of industry evolutions, set clear business goals to break the breakthrough barriers step by step and make investment decisions for the continuity of business in a digital transformation business environment.

2.2.3 Manufacturing Process

Zheng et al. (2021) presented various scientific literatures for a number of frameworks and references listing typical default manufacturing process in companies. There were several models used but two models are used in this study. The first model is the value chain analysis by Michael Porter which includes a list of activities undertaken by a company in order to deliver a product or service. The second model is the value reference model (VRM) (Kirikova et al. 2012), which provides three different levels, governance (strategical process), planning (statistical process) and execution (operational process). Development of new products, distribution network arrangement, integrated logistic making plans, internal logistics, manufacturing planning and scheduling, power management, quality assurance, network maintenance, customer relationship management and after-sales management are examples of typical manufacturing business operations identified from models (Zheng et al., 2021).

2.2.4 Supply Chain and Management

Supply chain and management is an important function that enables raw materials to reach the factory and finished products to reach the consumers. Suppliers, factories, distributors, retailers and consumers are connected through the logistics and transportation network process (Muduli & Barve, 2015; Biswal et al., 2019). Raw materials are transported from suppliers to the factory, and finished products are transported from factory to the distributors, then to retailers and finally consumers who buy them from retailers for consumption. This is a very difficult process to handle, but thanks to the digital revolution and the use of cloud computing, IoT and IoS, effective network and communication are now possible, ensuring that a product reaches its target location, market and, of course, the intended consumers.

For example, Mapai Transport and Logistics is a transport and logistics company that transports products from Lae to various highland provinces of PNG. Mapai has GPS tracking locations of trucks and containers. Customers (manufacturers like SP Brewery, Nestle PNG, Laga Industries, Coca Cola Amatil) are informed of their products in the container or customers themselves have access to this information through their phone or device in office. Also, Mapai uses a cloud-based accounting and freight management system that helps online invoicing and manages the movement and location of trailers, containers and trucks. Without logistics and transportation, products could not reach the intended market or customers. Smart phones, with the Internet, enable effective communication among manufacturers, transporters, distributors and retailers.

2.3   Circular Economy Concept

Circular economy concept (CEC) has developed from the influences of increased demand due to increase in population, resource scarcity, environmental protection and waste management and promotion of sustainable use of resources, with the main concern for future use of resources by future generations (Biswal et al., 2017; Muduli et al., 2021). Valavanidis (2018) referred circular economy concept as products are designed with the intention that after the end of product life cycle, the end product will become resources for others, closing the loops in industrial ecosystems, following sustainable methods and minimizing waste. The concept is simplified and represented by 3R's. 3R's represent the words, **Reduce, Reuse** and **Recycle**. The 3R's process guides product design at the factory to minimize waste generation (reduce), turn trash into a resource for reduction, reuse and recycling, and return to the manufacturer for remanufacturing. Therefore, this process can be concluded with the phrase 'closing loop or today's products are tomorrow's' raw materials (PWC). CEC has shifted from traditional linear concept of take-make and dispose as this model cannot support the balance and need of society. The Industry 4.0 has become a real enabler to both the manufacturer and CEC for the shift to CE from traditional economy. Figure 3 represents the circular economy concept.

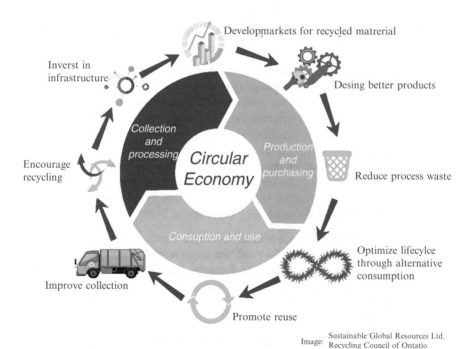

Fig. 3 illustrates circular economy concept. (Source: Sustainable Global Resources Ltd., Recycling Council of Ontario)

2.3.1   Importance (Benefits) of Circular Economy Concept

These advantages include increasing or extending the useful life of the product, lowering the risk of running out of resources, protecting the environment, meeting consumer demand through mass production, promoting efficient waste management, preparing for future consumption, making the manufacturing industry more valuable, and promoting business continuity or going concern. These benefits include increasing or extending the useful life of a product, lowering the risk of resources running out, protecting the environment, meeting consumer demand through mass production, getting ready for future consumption, increasing the value of a manufacturing business and keeping a business going or making sure it stays in business.

Product Useful Life The product useful life is extended by way of reuse and recycle. This can be achieved at the product design stage at the factory by using industry techniques including smart factories with CPS and AM. Extending the useful life of a product contributes to maximizing use of resources and managing resource scarcity.

Mitigate the Risk of Resource Scarcity The traditional linear economy, make-take-dispose, does not help in solving resource scarcity problem with the global growing population that demand more products each day. Global raw materials will be slowly depleted if an alternative action is not taken to address the resource scarcity. Industry 4.0 provides a solution to manufacturing industries to adopt CEC to manage and maximize use of scarce resources.

Promote Protection of Environment and Ecosystem Human life depends on environment and ecosystem. More exploitation into rain forest, land for agriculture, fishing and other natural resources puts human life in danger. Disposing of waste products into the environment will ruin the environment and ecosystem. Eventually, human life will be affected. Therefore, protecting the environment by promoting reduce, reuse and recycle principles means protecting human life. Again, this is enabled by Industry 4.0 at product design stage at the factory. Manufacturing industries promote awareness on environment conservation and governments like the European Parliament and the Chinese government by way of legislating the adoption of CEC promote the same thing (Valvanidis, 2018).

Mass Production in Meeting Consumer Demand Consumer demand is growing as population increases. With CEC and the support of Industry 4.0 techniques including smart factory with CPS, manufacturers can do mass production. Also, the extension of product life can help in managing demand. The reuse of second-hand products like clothes, cars and other appliance and devices can contribute to managing demand.

Promote Waste Management Reduce, reuse and recycle principles promote waste management system; therefore, waste disposed into the environment is minimized. Also, waste is turned into raw materials for remanufacturing and producing energy.

Prepare for Future Consumption The consumption of the past and of today is gone, but the consumption for the future is still there, and it is growing every day as the population grows. Industry 4.0 helps manufacturers make the best use of their resources when making things. Getting ready for the consumption by future generations is just as important as getting ready for today's consumption.

Increasing Value for Business According to the European Commission report (2014), the CEC can create an estimated gain of 600 billion euros for EU manufacturers. This is huge and good news for adopting CEC.

Going Concern for Manufacturing Business Businesses exist with a vision to continue the operations into the unforeseen future. Lack of change, innovation, adaption and adoption cannot solve the going concern for business. It is not an exception for manufacturing business. CEC promotes and solves the going concern for manufacturing business as it is empowered by Industry 4.0 which delivers innovation and opportunities to position businesses in the competitive business environment.

2.3.2 Industry 4.0 Enabling Manufacturing Industries in Adopting CEC in Both Developed and Developing Countries

Industry 4.0 techniques enable manufacturing industries to adopt circular economy concept. This is taking shape and widely accepted by manufacturers in both developed and developing countries. The European Parliament (2021) adopted a resolution on the new circular economy action plan directing more measures focusing on achieving carbon-neutral, sustainable environment, toxic free, and fully circular by 2050. Also, the parliament has tightened the recycling rules and binding targets for materials use and consumption by 2030. This puts more pressure on manufacturers to develop plans in implementing the action plan. The Chinese government has changed to circular economy concept and manufacturers have adopted CEC in production (Valvanidis, 2018). Zeng et al. (2017) explored the current state of Industry 4.0 adoption by Chinese manufacturers. Deloitte has investigated Swiss manufacturing companies positioning themselves in relation to the digital transformation and the opportunities Industry 4.0 provides. In some developing countries, most manufacturing companies are still working on the traditional linear economy model (Kumar et al. 2021a, b) and Industry 3.0.

3 Research Method

A literature review has been proposed as a research approach by a number of scholars because it can make substantial contributions to methodological, thematic and conceptual development in a variety of disciplines (Muduli et al., 2016; Snyder et al., 2019; Swain et al., 2021; Peter et al., 2022). It was decided to employ a qualitative method (QM) in this investigation, which was designed in two stages. The QM was chosen since it had previously been utilized in investigations by Nascimento et al. (2019) and Zheng et al. (2021), among others. The second stage involves collecting information from executives and operational specialists in selected manufacturing companies in PNG via telephone interviews and email questionnaires. The first stage involves reviewing literature from 40 journals, articles, research and news articles, and the second stage involves confirming the reviewed literature by collecting information from executives and operation specialists in selected manufacturing companies in PNG via telephone interviews and email questionnaires (Industry Focused Group or IFG). The information from the literature was evaluated using category analysis in order to acquire information that was relevant to the study's objectives. Since the study's goal is to present how Industry 4.0 techniques can support manufacturing companies in developing more promising circular economy concepts and encourage and even urge companies to adopt Industry 4.0 techniques and CEC, particularly in developing countries such as PNG, the approach taken in gathering information was more exploratory and descriptive (Nascimento et al., 2019).

3.1 Literature Review

The first stage of the filtering process consisted of categorizing 40 pieces of literature into four groups, as shown in Table 1.

The review studies were then broken down into three groups based on their goals, so that the ones that didn't fit into any of the groups could be thrown out.

- Exploring Industry 4.0 techniques and their applications in manufacturing industries for creating a more promising circular economy.
- Adoption of circular economy concept by manufacturing companies in both developed and developing countries.
- Barriers (challenges) in adopting Industry 4.0 that restrict the adoption of CEC.

The title, abstract and introductory section of each study were read in order to make the final selection during this process. At the second stage, any study that contained at least two of the three objectives set forth was accepted. At this point, 12 pieces of literature had been rejected, leaving another 12 to be considered.

The third stage consisted of filtering and selecting only those pieces of research that were close to meeting all three of the goals. Six pieces of literature were eliminated as a result of this approach, and six primary papers remained, which

Table 1 The categories of literatures reviewed

Journals	Articles	Industry study by PWC and Deloitte	News article	Total
33	3	2	2	40

Table 2 Six literature studies

Authors	Study type	Models and frameworks/approaches	Application
Rajput and Singh (2020)	Journal	Proposed new model (MILP) to minimize total costs and energy consumption of machines to set up industry 4.0 facility in achieving CEC	Manufacturing companies
Leunendonk (2019)	Article	Providing the descriptive approach of industry 4.0 and the trend of digitalization	Manufacturing companies
Deloitte	Research	Exploring industry 4.0 challenges and providing solutions	Manufacturing companies
Nascimento et al. (2019)	Journal	Theoretical and systematic approach to affirm the need for adoption of CEC by manufacturing companies	Manufacturing companies
PWC	Research	Providing descriptive approach highlighting the trend and solutions provided by industry 4.0 and CEC	Manufacturing companies
Zheng et al. (2021)	Journal	This paper provides a systematic approach and setting framework for adoption of industry 4.0 technologies	Manufacturing companies

were then read and examined. Several literature studies, including six in Table 2, have focused on Industry 4.0 strategies that are enabling manufacturers in both developed and developing countries to implement CEC. The barriers have also been identified, but the solutions supplied by both Industry 4.0 and CEC can inspire firms to break down those barriers and gradually adopt CEC technology. Applications of Industry 4.0 technologies are shown in Table 3 for companies in the manufacturing industry. Table 4 lists the things that are keeping more people from using Industry 4.0 and CEC.

3.1.1 Exploring Industry 4.0 Techniques and Their Applications in Manufacturing Industries for Creating a More Promising Circular Economy

The application of Industry 4.0 techniques in manufacturing companies that enable the adoption of circular economy concept is summarized as follows.

3.1.2 Adoption of Circular Economy Concept by Manufacturing Companies in Both Developed and Developing Countries

According to the literature studies, German and other European countries are leading in developing Industry 4.0 techniques and adopting CEC (European Parliament 2021, Deloitte). In the Asian region, China is firm in adopting CEC (Valvanidis, 2018). The United States is focusing on the CEC as well. North and South Korean

Table 3 Application of Industry 4.0 techniques in manufacturing companies

Industry 4.0 technique	Applications (description)	Circular economy concept
Cyber-physical systems (CPS)	Smart product development, scheduling, control, design and improvement. Risk management, integration and automation of supply chain and service oriented	Sharing economy, maintenance, repair, reduce, reuse, recycling
Internet of things (IoT)	Enables communication, interaction and cooperation among sensors, smart phones and machines to work out a solution	Sharing economy, maintenance, repair, recycling
Internet of data and analytics (IoDA)	Support in collecting large data and filter, capture, report for improvement	Sharing economy, maintenance, repair, reduce, reuse, recycling
Smart factory (SF)	Connects physical world and virtual world. The CPS communicate over the IoT and assist people and machines in the execution of their tasks	Sharing economy, maintenance, repair, reduce, reuse, recycling
Internet of services (IoS)	Connects supply chain including suppliers, manufacturers, distributors, retailers and customers. Helps manufacturers to collect information and respond to customer demand	Sharing economy, maintenance, repair, reduce, reuse, recycling
Cloud computing (CC)	Enables storing all data and programs for safe and easy access	Sharing economy, maintenance, repair, reduce, reuse, recycling
Blockchain (BC)	Smart purchase and supply of products, tracking materials and virtualization	Sharing economy, maintenance, repair, reuse, recycling
Artificial intelligence (AI) and robotics	Made to do and act like human with more efficiency and accuracy. Optimize production	Reduce, reuse, recycling, remanufacturing
Additive manufacturing (M) and 3D printing	Increase product customization, reduce cost of productions, flexible to adapt	Sharing economy, maintenance, repair, recycling

Adopted from Zheng et al. (2021), Nascimento et al. (2019)

Table 4 Common challenges in adopting Industry 4.0

Industry 4.0 challenge	Description
Data and information security	Online data integration and network, sharing information opens gap for losing data and business information
Capital costs	Huge spending (cost) in setting up industry 4.0 facilities is grave concern. Lack of leadership and knowledge on the industry discourages investment
Skills, training and innovation	Lack of relevant skills, training and innovation at managerial and specialist with key personnel

companies, especially SME, are moving at slowly towards CEC (Zheng et al., 2021). Most developing countries are still working with the Industry 3.0 or traditional linear model (Kumar et al., 2021a, b).

3.1.3 Barriers (Challenges) in Adopting Industry 4.0 That Restricts the Adoption of CEC

The common Barriers found in the literature that restrict and delay the adoption of Industry 4.0 and subsequent adoption of CEC are summarized in Table 4.

3.1.4 Industry Focus Group (IFG)

The following questions were formed to get the views and current positions of manufacturers in PNG. IFG includes four executive managers, four factory specialist engineers and two supply chain managers, and they agreed to provide information on the following questions. Six of them were interviewed by telephone and four have completed the questionnaires and returned by email. This commenced on the 24th of May 2021 and competed on the third of June 2021.

- Has your manufacturing company adopted the concept of reduce, reuse and recycling process (methods)?
- Is your manufacturing company using Industry 4.0 techniques or digital technologies at your factory for production? If the answer was No and below satisfactory level (less than 50% of satisfactory), the four common barriers (challenges) that include information security risk, capital costs, skills, training and innovation and lack of government direction and action plan were asked and they have provided their responses. They were summarized in Table 6.

Table 5 provides the respondents to indicate the current practice for the concept of reduce, reuse and recycle (3Rs) in their respective organizations. The respondents say, no, which means they do not practise any of the 3Rs and yes, means they practice. The total number is provided under each category. Most of them (90%) responded by saying they produce to sell and when the product is sold, the activity closed, and they order new raw materials for production.

All executives and specialists responded that they are using digital technologies but not like Industry 4.0. They said, mass production is done to meet demand but when product is sold, they move on to purchase new raw materials for production. Therefore, there is no reduce, reuse and recycling for the selected manufacturers. The recycle as shown in Table 5 with three respondents was not done by the manufacturing companies but by different companies for different uses, not by the same manufacturing company. Reduce showing in Table 5 with three respondents is relating to waste from production, not product waste. Table 6 shows challenges (barriers) that the selected manufacturers are facing that prevents the adoption of Industry 4.0 and circular economy concept. Each person indicates by saying yes or

Table 5 Results of interviews and questionnaires

Respondent	Reduce	Reuse	Recycle
Yes	10 (100%)	3 (30%)	3 (30%)
No	0	7 (70%)	7 (70%)
Total	10	10	10
%	100% production waste management	30% but not all	30% but for only one product and the rest by others

Table 6 Responses regarding barriers of adopting Industry 4.0

Respondent	Information and data risk	Capital cost	Skills, training and innovation	Lack of government initiatives and action plan
Yes	0	0	0	0
No	10	10	10	10
Total	10	10	10	10
%	100%	100%	100%	100%

no under each category and the total number is written down. No means the respondents agree that the listed challenges prevent their organization from adopting Industry 4.0 and CEC.

4   Results and Discussions

The studies on Industry 4.0 are mostly focused on manufacturing companies and promoting CEC. More research is required to close the gaps especially barriers such as data and information security for all manufacturing companies to adopt CEC with the opportunity provided by Industry 4.0.

4.1   Results from Literature Review

In one literature review, the MILP (Mixed-Integer Linear Programming) model for achieving circular economy was proposed, while others provide frameworks and descriptive and systematic approaches that emphasize the process and values and identify barriers that manufacturers can overcome in the interest of the company. Listed below are the most prevalent Industry 4.0 strategies that have been offered in the literature, together with their specific functions in achieving the circular economy concept. CPS, IoT, IoDA, SF, IoS, CC, BC, AM, AI and robotics are some of the technologies involved. Each function contributes to the achievement of the overall goal, which is to promote and realize CEC, which includes the sharing economy, preventative maintenance, reuse, recycling, remanufacturing, waste management and environmental protection, among other things.

Industrialized countries are at the forefront of the development and acceptance of Industry 4.0 and the CEC (Common European Framework). These countries include members of the European Union, China and the United States. Korea, for example, is one of the Asian countries that has adopted the trend. Most developing countries haven't adopted Industry 4.0 and CEC yet. They still use the linear economic paradigm of 'make, use, and dispose of', as well as Industry 3.0 (the third industry evolution).

For the adoption of Industry 4.0 and CEC, there were three common obstacles to overcome. Data and information security, capital costs and skills, training and innovation were the three topics covered. Because of the online data and information sharing provided by Industry 4.0 techniques such as the Internet of Things, the Internet of Things Services, cloud computing and big data, manufacturing companies are hesitant to adopt Industry 4.0 and CEC for fear of losing valuable company data and information to competitors and others. Another obstacle that results from a lack of inventive leadership and the inability to obtain accurate market knowledge for development is the high cost of capital. In order to successfully implement Industry 4.0 and CEC, skills, training and innovation must be available. A lack of these critical change components is a significant problem.

4.2   Results from Industry Focused Group (IFG) in PNG

According to the results of the study, reduce, reuse and recycling are not actively done by the industrial companies that were chosen. According to the respondents, production waste is well handled, but product waste after the product life cycle is not. In the case of production waste management, every single one of the ten respondents (100%) responded. IFG stated that mass production is carried out solely for the purpose of selling the goods, implying that once the product reaches the consumer, the end result is of less importance to the makers. They place new orders for raw materials in preparation for new production. One in three people (30%) said that their product package or can is recyclable, but another company buys it to make a new product rather than them. Only bottles are collected and recycled for remanufacturing and public reuse. Seven (70%) of those polled stated that their organization does not practise reuse or recycling.

With regard to challenges, the IFG responded that their organizations face the same issues as everyone else: information and data security, high capital costs for setting up Industry 4.0 facilities, a scarcity of skilled workers, inadequate training and a lack of government initiatives and an action plan. Each category of challenge received a unanimous 'no' response from all ten participants (100%).

4.3 *Discussions*

Industry 4.0 is a true enabler in terms of lowering costs and increasing value throughout the supply chain, from the manufacturer to the consumer. All Industry 4.0 techniques were designed on the basis of design principles in order to meet the needs of consumers and the CEC. Technology for Industry 4.0 is still in the early stages of development. For example, one of the studies included in this analysis presented a different model (MILP). All countries are concerned about CEC (reduce, reuse, recycle and remanufacture), but developed countries and a few other developing countries are leading the way with urgency as their leadership in understanding the past, present and future development trends, information technology infrastructure, skills and knowledge depletion. Continued study, public awareness and progress in the nations that have embraced Industry 4.0 and CEC will motivate others to manage and overcome the barriers that stand in the way of adopting Industry 4.0 and CEC in the long run, as CEC is a worldwide concern. Furthermore, manufacturing companies as a business want to leave and keep going into the future, not be forced to stop doing what they are doing now.

PNG, which is a poor country, is lagging behind in terms of implementing Industry 4.0 and CEC technologies. Reduce, reuse and recycle results from the IFG reveal that PNG manufacturers have done less work to attain the level of Industry 4.0 and the CEC than other countries. The study confirms and concludes that the classic linear economic model is used by the majority of manufacturers in Papua New Guinea. Only bottles are recycled and utilized by the general population. This results in a lower contribution to the CEC. Other companies reuse cans, but this isn't a project started by the people who made the original products or their subsidiaries.

The barriers to Industry 4.0 continue to be a significant concern for PNG, where the cost of doing business is high, including the capital investment required to implement Industry 4.0 processes and the unreliability of electricity. Despite the fact that policies are already in existence and there has been much discussion on the matter, the lack of government direction and action plans in IT infrastructure and policy enforcement continues to be a problem. Businesses continue to face significant risks when it comes to data and information security. Manufacturing businesses are concerned that competitors or anyone else would have access to company information, which might pose a danger to their products and businesses in the market. In the manufacturing industry, not having enough training, skills and ideas is a bigger problem than in other jobs.

The future of PNG manufacturing is in jeopardy because other countries, such as China, are already implementing Industry 4.0 technology in their manufacturing processes and implementing CEC. The Chinese products available on the shelves of PNG wholesalers and retailers are far less expensive than locally made goods. Some factories may relocate to other countries and only bring in finished goods for resale in the United States. Additionally, the raw materials and environment of PNG are under threat as a result of the current tendency towards a linear economy. This is an economic challenge that the government must solve as soon as possible by examining existing policies and legislation and modifying them to better meet current trends for rapid implementation; otherwise PNG will be further left behind in the global economy. Businesses exist in order to carry on their activities into the future, but those that do not adapt and alter in response to change will be forced to close their doors. The manufacturing sector in Papua New Guinea cannot afford to wait for anyone, not even the government, to reform in order to respect corporate duties and operate in an environmentally sustainable manner. It is very important for them to keep working if they respond to changes, adapt and adopt the right way.

5 Limitations and Recommendations

This study has a number of limitations, some of which are as follows:

- The 40 literature review papers that were employed in this study were not sufficient to gain a more comprehensive understanding of the issue in order to answer the query. In addition, there was not enough time to read through all of the articles in depth.
- IFG's focus was solely on food manufacturing businesses. This study does not include all of the manufacturers in question. In addition, time constraints made it impossible to reach out to all of those who needed help.

If successful, the study will be the first of its kind for PNG manufacturing enterprises and the industry. It was enlightening to learn about the trend towards industrialization, which is being driven by digitalization. Furthermore, it was surprising to find that underdeveloped countries such as PNG are not eager to adapt to change and implement Industry 4.0 and CEC initiatives. Due to rising demand and scarcity of resources, there is an imbalance between economic principles in practice at the present time. As a result, the following recommendations for consideration and action are made:

- Further investigation into the impediments to Industry 4.0 is needed to close the gap. This will finish the transition to Industry 4.0 and give everyone a way to use CEC.
- Industry 4.0 and the CEC are being closely scrutinized and studied by industrialized countries in order to prepare for their own adoption.
- In the same way that Germany, the European Union and China took ownership of CEC as an economic concern, governments in any country may do the same in order to face the challenges and opportunities presented by digitization.
- Companies in the manufacturing sector in Papua New Guinea should take the initiative as a firm to uphold corporate principles and ethical standards, be innovative, and aim to adopt Industry 4.0 and CEC practices.

- When it comes to information technology, the manufacturing industry and related existing legislation, the PNG government can evaluate and alter existing regulations to enable and transform manufacturing businesses in a challenging business climate.

6 Conclusion

Building on six customer (consumer) and CEC-focused fundamental concepts, Industry 4.0 will become a real enabler, providing solutions and adding value to entities in the supply chain, the environment, government and economies of any country in which it operates. As a result of the current trend of increased consumer demand and behaviour combined with resource constraints, the CEC is required. Additionally, environmental degradation and the depletion of basic materials are concerns. The study about Industrial 4.0 techniques (components) continues to be a challenge since it is difficult to define and conclude the necessary Industry 4.0 techniques that will close the loop, because today's waste products are tomorrow's raw materials. Further study on data and information security is required in order to close the gap between now and the deployment of Industry 4.0. Other roadblocks, such as capital expenses, skills, training and innovation, will be addressed as needed by the company in order to ensure the existence and continuity of the enterprise. Any manufacturing company that fails to adapt to the changing business climate in the digitalized era will be left behind or pushed out of existence. PNG's government initiatives and action plan on IT development and CEC are still a big problem.

References

Biswal, J. N., Muduli, K., & Satapathy, S. (2017). Critical analysis of drivers and barriers of sustainable supply chain management in Indian thermal sector. *International Journal of Procurement Management, 10*(4), 411–430.

Biswal, J. N., Muduli, K., Satapathy, S., & Yadav, D. K. (2019). A TISM based study of SSCM enablers: An Indian coal- fired thermal power plant perspective. *International Journal of System Assurance Engineering and Management, 10*, 126–141.

Dong, X., Yu, J., Luo, Y., Chen, Y., Xue, G., & Li, M. (2014). Achieving an effective, scalable and privacy-preserving data sharing service in cloud computing. *Computers & Security, 42*, 151–164.

Esmaeilian, B., Behdad, S., & Wang, B. (2016). The evolution and future of manufacturing: A review. *Journal of Manufacturing Systems, 39*, 79–100.

European Parliament. (2021, March). Economy update. https://www.europarl.europa.eu/news/en/headlines/economy/20151201STO05603/circular-economy-definition-importance-and-benefits

Ghobakhloo, M. (2018). The future of manufacturing industry: A strategic roadmap toward industry 4.0. *Journal of Manufacturing Technology Management.*

Ivanov, D., Dolgui, A., Sokolov, B., Werner, F., & Ivanova, M. (2016). A dynamic model and an algorithm for short-term supply chain scheduling in the smart factory industry 4.0. *International Journal of Production Research, 54*(2), 386–402.

Joshi, S., Sharma, M., Das, R. P., Muduli, K., Raut, R., Narkhede, B. E., et al. (2022). Assessing effectiveness of humanitarian activities against COVID-19 disruption: The role of blockchain-enabled digital humanitarian network (BT-DHN). *Sustainability, 14*(3), 1904.

Kang, H., Lee, J., Choi, S., Kim, H., Park, J., Son, J., Kim, B., & Noh, S. (2016). Smart manufacturing: past research, present findings, and future directions. *International Journal of Precision Engineering and Manufacturing – Green Technology, 3*(1), 111–128.

Kirikova, M., Buchmann, R., & Costin, R. A. (2012, September). Joint use of SCOR and VRM. In *International conference on business informatics research* (pp. 111–125). Springer.

Kumar, P., Singh, R. K., & Kumar, V. (2021a). Managing supply chains for sustainable operations in the era of industry 4.0 and circular economy: Analysis of barriers. *Resources, Conservation and Recycling, 164*, 105215.

Kumar, S., Raut, R. D., Narwane, V. S., Narkhede, B. E., & Muduli, K. (2021b). Implementation barriers of smart technology in Indian Sustainable Warehouse by using a Delphi-ISM-ANP approach. *International Journal of Productivity and Performance Management, 71*(3), 696–721.

Lee, J., Bagheri, B., & Kao, H. A. (2015). A cyber-physical systems architecture for industry 4.0-based manufacturing systems. *Manufacturing Letters, 3*, 18–23.

Li, Z., Wang, Y., & Wang, K. S. (2017). Intelligent predictive maintenance for fault diagnosis and prognosis in machine centers: Industry 4.0 scenario. *Advances in Manufacturing, 5*(4), 377–387.

Lom, M., Pribyl, O., & Svitek, M. (2016). Industry 4.0 as a part of smart cities. In *Smart Cities Symposium Prague (SCSP)* (pp. 1–6).

Luenendonk. (2019). https://www.cleverism.com/industry-4-0

Mastos, T. D., Nizamis, A., Terzi, S., Gkortzis, D., Papadopoulos, A., Tsagkalidis, N., et al. (2021). Introducing an application of an industry 4.0 solution for circular supply chain management. *Journal of Cleaner Production, 300*, 126886.

Monostori, L., Kádár, B., Bauernhansl, T., Kondoh, S., Kumara, S., Reinhart, G., et al. (2016). Cyber-physical systems in manufacturing. *CIRP Annals, 65*(2), 621–641.

Muduli, K., & Barve, A. (2015). Analysis of critical activities for GSCM implementation in mining supply chains in India using fuzzy analytical hierarchy process. *International Journal of Business Excellence, 8*(6), 767–797.

Muduli, K., Barve, A., Tripathy, S., & Biswal, J. N. (2016). Green practices adopted by the mining supply chains in India: A case study. *International Journal of Environment and Sustainable Development, 15*(2), 159–182.

Muduli, K., Kusi-Sarpong, S., Yadav, D. K., Gupta, H., & Jabbour, C. J. C. (2021). An original assessment of the influence of soft dimensions on implementation of sustainability practices: Implications for the thermal energy sector in fast growing economies. *Operations Management Research, 14*(3), 337–358.

Muduli, K., Raut, R., Narkhede, B. E., & Shee, H. (2022). Blockchain technology for enhancing supply chain performance and reducing the threats arising from the COVID-19 pandemic. *Sustainability, 14*(6), 3290.

Nascimento, D. L. M., Alencastro, V., Quelhas, O. L. G., Caiado, R. G. G., Garza-Reyes, J. A., Rocha-Lona, L., & Tortorella, G. (2019). Exploring industry 4.0 technologies to enable circular economy practices in a manufacturing context: A business model proposal. *Journal of Manufacturing Technology Management., 30*(3), 607–627.

Oztemel, E., & Gursev, S. (2020). Literature review of industry 4.0 and related technologies. *Journal of Intelligent Manufacturing, 31*(1), 127–182.

Parajuly, K., & Wenzel, H. (2017). Potential for circular economy in household WEEE management. *Journal of Cleaner Production, 151*, 272–285.

Peter, O., Swain, S., Muduli, K., & Ramasamy, A. (2022). IoT in combating COVID-19 pandemics: lessons for developing countries. In *Assessing COVID-19 and other pandemics and epidemics using computational modelling and data analysis* (pp. 113–131). Springer.

Rajput, S., & Singh, S. P. (2020). Industry 4.0 model for circular economy and cleaner production. *Journal of Cleaner Production, 277*, 123853.

Russell, S. J., & Norvig, P. (2016). *Artificial intelligence: A modern approach.*

Sikorski, J. J., Haughton, J., & Kraft, M. (2017). Blockchain technology in the chemical industry: Machine-to-machine electricity market. *Applied Energy, 195*, 234–246.

Snyder, L. V., Atan, Z., Peng, P., Rong, Y., Schmitt, A. J., & Sinsoysal, B. (2019). OR/MS models for supply chain disruptions: A review. *IIE Transactions, 48*(2), 89–109.

Sousa-Zomer, T. T., & Miguel, P. A. C. (2018). Sustainable business models as an innovation strategy in the water sector: an empirical investigation of a sustainable product-service system. *Journal of Cleaner Production, 171*, S119–S129.

Swain, S., Peter, O., Adimuthu, R., & Muduli, K. (2021). Blockchain technology for limiting the impact of pandemic: Challenges and prospects. In *Computational modeling and data analysis in COVID-19 research* (pp. 165–186). CRC Press.

Thoben, K. D., Wiesner, S., & Wuest, T. (2017). "Industrie 4.0" and smart manufacturing-a review of research issues and application examples. *International Journal of Automation Technology, 11*(1), 4–16.

Trappey, A. J., Trappey, C. V., Govindarajan, U. H., Sun, J. J., & Chuang, A. C. (2016). A review of technology standards and patent portfolios for enabling cyber-physical systems in advanced manufacturing. *IEEE Access, 4*, 7356–7382.

Valavanidis, A. (2018, July). Concept and practice of the circular economy. *Scientific Reviews*, 1–30.

Vera-Baquero, A., Colomo-Palacios, R., & Molloy, O. (2014). Towards a process to guide big data based decision support systems for business processes. *Procedia Technology, 16*, 11–21.

Wamba, S. F., Akter, S., Edwards, A., Chopin, G., & Gnanzou, D. (2015). How 'big data' can make big impact: Findings from a systematic review and a longitudinal case study. *International Journal of Production Economics, 165*, 234–246.

Wan, J., Tang, S., Shu, Z., Li, D., Wang, S., Imran, M., & Vasilakos, A. V. (2016a). Software-defined industrial internet of things in the context of industry 4.0. *IEEE Sensors Journal, 16*(20), 7373–7380.

Wan, J., Yi, M., Li, D., Zhang, C., Wang, S., & Zhou, K. (2016b). Mobile services for customization manufacturing systems: An example of industry 4.0. *IEEE Access, 4*, 8977–8986.

Zeng, H., Chen, X., Xiao, X., & Zhou, Z. (2017). Institutional pressures, sustainable supply chain management, and circular economy capability: Empirical evidence from Chinese eco-industrial park firms. *Journal of Cleaner Production, 155*, 54–65.

Zhang, L., Luo, Y., Tao, F., Li, B. H., Ren, L., Zhang, X., et al. (2014). Cloud manufacturing: A new manufacturing paradigm. *Enterprise Information Systems, 8*(2), 167–187.

Zheng, T., Ardolino, M., Bacchetti, A., & Perona, M. (2021). The applications of industry 4.0 technologies in manufacturing context: A systematic literature review. *International Journal of Production Research, 59*(6), 1922–1954.

Smart Technologies Interventions for Sustainable Agri-Food Supply Chain

Suyash Manoram and Anupama Panghal

1 Introduction

Global demand for food products is increasing with the rise in population and customer's tastes and variance. The supply chain becomes the crucial part to meet the rising demand for food products. Issues such as need for environmental and economical sustainability have further strengthened the reason to make the supply chain more sustainable and better developed. With the introduction of smart technologies, there can be an enhancement in the performance of the supply chain. Smart technologies may be adopted in the sustainable agriculture supply chain to provide high value to the farmers (by strengthening their position), to industry (by helping them to efficiently use the raw material) and also to consumer (to have their desired product sitting in the remote area at an optimum cost).

In this era, sustainability is the most important criteria for the design and operation of the supply chain. Sustainability-oriented focus allows the supply chain to target the environment concerned customer and improve the performance of the supply chain. Every individual supply chain occupies a very specific niche in this world; thus for the health and survival of every supply chain and every individual, the health of the environment is important (Chopra & Meindl, 2016). For making the supply chain sustainable, one should expand the goals of the supply chain without focusing on an individual's profit. These goals can be achieved by focusing on the use of non-renewable resources which reduce environmental haz-

S. Manoram · A. Panghal (✉)
Department of Food Business Management and Entrepreneurship Development, National Institute of Food Technology Entrepreneurship and Management (NIFTEM), Kundli, Haryana, India

ards. A supply chain should not be concerned only with increasing its surplus rather it should be focused on the upliftment of society, reduce its negative impact on the environment and have a responsibility towards its behaviour on the issues like child labour adulteration, pollution, etc. (Santiteerakul et al., 2020). Supply chain management brings out many concepts to make the supply chain sustainable by performing functions suitable as per environmental concern like reducing waste, reducing carbon emission, etc. It also keeps social issues in mind such as the working condition of the supplier and resource procurement in ethical and legal ways. Supply chain management focuses on the economic aspect by buying goods from the local supplier. The process of being sustainable varies from organization to organization; some focuses on environmental concern, some have social priorities, while others may focus on earning profits (Joshi et al., 2020).

Technology changes people's attitude and approach regarding information, communication, manufacturing and interaction in this period. Performance is greatly enhanced in each process like procurement process, manufacturing process, distribution process, etc., of the organization using smart technology specifically in supply chain system resulting in varied improvements of desired targets and creating value for the whole supply chain (Seuring, 2012). The technologies have a great impact in converting the supply chain into a sustainable supply chain by taking concern of both environmental and social parameters thus improving the environment by reducing wastes, reducing the overexploitation of natural resources, reducing the emission of hazardous gases, using renewable resources for carrying out operations, etc. On the other hand, on societal aspects, it removes product adulteration, increases consumer awareness, provides information regarding the quality of product, providing information access about the location of the product, detecting the originality of the product, removing the drudgery of labour, etc.

Nowadays every organization focuses on beating their competitors by gaining the technological advantages for the operations performed by this organization while having concern on environment's health thereby converting their supply chain into a sustainable supply chain. Concerning the covid-19 pandemic, there is a global issue of sustainability affecting various parameters of a supply chain in every organization (Ben-Daya et al., 2019). The introduction of new and smart technologies like artificial intelligence (AI), blockchain, Internet of Things, sensors, wearable technologies, 3D food printing, etc. enhances the organization status by targeting the consumers in terms of providing a quality product, proper tracking information, products at low cost, less lead time, secure environment, etc.

2   Sustainable Agri-Food Supply Chain

When a supply chain works on the process of reducing the negative effect on the environment, improving the social welfares and improving its life, then that supply chain is called the sustainable supply chain. Nowadays many stakeholders demand from the management to make a sustainable supply chain for diversifying and enhancing the life of the supply chain. There is pressure from all over the globe to improve the quality of product and enhance the working condition of labour: health and safety of workers, sustainability of the environment, procurement strategy, buyer-supplier relationship and the life cycle of product (Nasiri et al., 2020a, b). Being a sustainable supply chain manager, priority is given to make the organization socially responsible and become sustainable by keeping the concern of protecting the environment, biodiversity, natural resources and landscapes.

In the present scenario, organizations are focusing more towards a sustainable supply chain as it provides the access to the new market where the environment-concerned people buy the products that match with their ideologies and the very same consumers influence others and hereby promote the product. It is a global trend to protect the environment, save biodiversity and let our inheritance enjoy the beautiful climate and use natural resources. Sustainability also gives competitive advantages as it touches the emotion of people and makes them aware that we are also trying to save the climate and mother earth. A sustainable supply chain brings value for the customer as the maintenance required for the product is low, and it also improves the health condition of the people and most importantly it reduces the overuse of the resources thus protecting the environment.

A sustainable supply chain has more advantages even though it has some barrier to come into existence. Cost is the main driver in the transformation of the supply chain into a sustainable supply chain. A supply chain is basically the collaboration of producer-manufacturer-distributor-retailer-customer so when a sustainable chain comes into action it reduces the current profit of a single partner but ultimately increases the life and profit of the supply chain. Parties seeking immediate profit try misaligned practices which are not good for the life extension of the supply chain, and thus the ultimate goal of transforming the supply chain into a sustainable one fails to come into existence. Ignorance is valid for both the supply chain manager and the consumer whose preference lies between exchanging low-cost product without taking into account the environmental factors. Consumers also show reluctance to buy the product from a sustainable supply chain due to the cost factor. Furthermore, few organizations ignore the condition of the environment, they just want to make a profit, through the overexploitation of natural resources and ultimately, they fail to function as a sustainable supply chain. For transforming a supply chain into a sustainable supply chain, an organization requires a huge sum of investment from their shareholder to introduce smart technologies. The investment is used to change the manufacturing process to modify or buy new vehicles for transportation purpose, to increase coordination for reducing the lead time or to provide value to the customer. It is one of the major hurdles as with high investment, most organizations are reluctant to introduce smart technologies in their systems.

3   Role of Smart Technologies in the Sustainable Agri-Food Supply Chain

For improving the quality of the supply chain several new technologies come into action, namely, artificial intelligence, the Internet of Things, blockchain, robotics, sensors, wearable technologies, etc. The role of these smart technologies is discussed further in this section:

3.1   Role of Artificial Intelligence in the Sustainable Agri-Food Supply Chain

AI deals with the group of technologies that are supposed to be done by human intelligence. It is the computer system that helps in object identification, resource allocation, tracking, failure prediction, etc. It is used to control the robot on its own or under human direction. The main purpose of it is to enhance human performance (Abbasi & Nilsson, 2012). It helps in facilitating at every level – for example, for an agricultural supplier, it facilitates to make plant disease-free; for the manufacturer, it can assist to detect the damaged raw material; for distributor, it promotes better coordination; for a retailer, it streamlines in tracking and for the customer, it helps to determine the best condition of the product before consumption. For making a sustainable supply chain, there are risks at each level, and dealing with the same helps in increasing its performance efficiently. AI makes the supply chain flexible enabling them to tackle the uncertain demand, implied demand uncertainty, and maintain stability while enhancing its performance and longevity. It promotes improving the bonding between the supply chain partners, keeping in view the performance of the outsourcing partner. It assists in enhancing the whole supply chain surplus, creating value for all the partners involved, including customer, thus maintaining their interest in the organization.

3.2   Role of Internet of Things in Sustainable Agri-Food Supply Chain

As the name suggests, IoT works by the Internet. For functioning properly, it seeks the assistance of RFID (Radio Frequency Identification) technology and GPS (Global Positioning System), sensory devices and the Internet. It mainly helps in the storage of information, transport navigation, billing, etc. (Asaf, 2020). Its speciality of work in the supply chain is to reduce the bullwhip effect by correct information transferring. It helps in the reduction of risk level within the supply chain and identifying the partner who is not working properly in the supply chain. The bullwhip effect causes huge fluctuation of demand for the supplier, and using the Internet of Things like RFID and GPS reduces this effect indirectly thus saving the climate.

3.3   Role of Blockchain in the Sustainable Agri-Food Supply Chain

Blockchain is used to increase the transparency in the supply chain by providing the participants with the power to record price, date, location, quality, etc. It eliminates the intermediaries by implementing the exchange of food products and transactions between different groups. The traceability of the products in the supply chain can be easily maintained through this technology. Outsourcing companies get a tremendous advantage with their overall review of the functions and get a competitive advantage over competitors. Blockchain aids in proper data sharing, reducing paperwork and administrative work and thus enhancing public trust and the company's stakeholder. Companies can easily track their inbound and outbound logistics making it feasible to coordinate with other participants along the supply chain and reduce the inbound logistic cost and time for the procurement process by the companies. Reduction of fraudulent activities for goods having high order and providing access to information to every party enhances the value and performance of the supply chain. It helps to reduce the drudgery of the labour as manual paperwork consumes lots of time and energy (Mostafa, 2019).

3.4   Role of Robotics in Sustainable Agri-Business Supply Chain

For expanding their business, an organization has to focus on the effectiveness of their supply chain, which can be achieved by following the principle of being cost-efficient, being responsive and converting the supply chain into a value chain (Saurabh & Dey, 2020). These goals cannot be achieved by manual labour only since there are human tendencies that we can't change. Those tendencies may be boredom, ignorance, mistakes, incentive gaining approach, over-expectation, etc. Thus, to be on the top of the competition and to achieve a goal, an organization has to introduce robotics in the working field. These robots are introduced either to assist labour or to do work in place of them. Robots are used to enhance human capabilities by saving time and reducing labour wages, and they are accurate in their calculation. If there is some boring or repeating task, then robots are most useful because during that same time human can work on some productive tasks. Nowadays robots are used at all the functional point of the supply chain, be it supplier, manufacturer, distribution centre or retailer. At every functional point of a supply, it is found that robotics helps in doing things efficiently, which is time saving and brings profit to the organization. It also helps in the reduction of waste, upliftment of the working condition for the labourer and generation of more and more revenue.

3.5 Role of Sensors in Sustainable Agri-Food Supply Chain

Smart sensors are the devices which convert physical things into digital medium to bring holistic approach in supply chain and convert supply chain into the value chain. Sensors find their role in three different platforms such as aerial, ground-based mobile and stationary systems (Quinnell, 2019). It aids the supply chain manager in keeping the information of inventory, tracking of the transportation, machinery, keeping the record of purchased good, buying pattern of the customer and maintaining ambience in cold storage, warehouses, distribution centre, retailer store, etc. These have been involved to assist in the number of processes which are (i) forecasting of demand, (ii) design collaboration in the sourcing process, (iii) increasing efficiency of the operational process, (iv) taking manufacturing decision, (v) decreasing the cost of production, (vi) saving time and (vii) enhancing the performance of the labour. Collaboration between supplier, manufacturer and distributor is greatly enhanced through digital advancement providing access to end-to-end visibility. Nowadays organizations implement these technologies to anticipate future demands and required services (Duong, 2020).

3.6 Role of Wearable Technologies in Sustainable Agri-Food Supply Chain

Wearable technologies consist of those elements which can be dislocated from one place to another. These wearable technologies come as easy to handle device which assists in the proper functioning of the supply chain by reducing the human effort. These devices assist in tracking and gaining information regarding inbound and outbound logistics. The main examples of wearing devices are watches and belt. The major role of these technologies in building the supply chain is to maintain the coordination between the partners by sharing information from the different locations. Manual workers are benefitted from multi-tasks, reducing time and increasing efficiency which reduces labour cost and drudgery. The working environment can be improved by detecting each parameter of the work, thus boosting the status of the health of the work and its workers. A positive image along with emotional advantages is established in the organization giving tough competition to its competitors.

3.7 Role of 3D Food Printing in Agri-Food Sustainable Supply Chain

The process of 3D printing uses a three-dimensional digital model to create a physical object by incorporating many thin layer substances in succession. It finds numerous applications in this field, and with a specific audience, it has set out scope for opportunities and advantages, providing customized consumer needs to attract them, thereby reducing internal costs and increasing efficiency; automated food production processes which is simple, time convenient and easy to operate thus reducing lead time; and innovation of range of structures and shapes, utilizing substitute resources and improving sustainability through food waste reduction which is one of the enormous tasks of the organization (Yang et al., 2017; Dankar et al., 2018). Every organization wishes to bring this technology to work to gain competitive advantages among its competitors by managing the procurement from the local market instead of sourcing it from a remote location. This step helps to uplift the local society while that organization gains emotional advantages in the market. It also benefits the organization by reducing its inbound logistic cost. Workers are benefitted from the removal of the burden by reduction of complexities and enhancing their performance. Greening the process saves the environment, reduces waste, reduces overproduction and makes the supply chain cost-efficient as well as responsive. Companies' demand can be easily met with the application of this technology, reducing the inventory cost, inbound logistic cost, warehouse cost, etc. (Voon et al., 2019).

4   Review of Literature

The introduction of smart technologies enhances the profit for the supply chain by reusing and reducing waste (Sardar et al., 2021). While smart technologies provide transparency and value to the customer, it also shifts the agri-food supply chain towards sustainability. Since management of agri-food supply chain is not an easy job to do, researchers came up with blockchain as a smart technology for safety and symmetry of information being accessed (Saunila et al., 2019). Also, the solicitation of such technology can lead to various improvements in agri-food supply chain competence and management of quality, e.g.;cloud computing and machine learning. With the introduction of new technologies, holding cost of inventory in the supply chain is reduced as there is an increase in storing duration with the help of smart technologies (Mostafa et al., 2019). The researchers found a major loophole in using technologies in the agri-food supply chain which is data tampering. To handle this loophole, the use of blockchain has been suggested. Blockchain has evolved as a rising technology for keeping track of the supply chain because of non-trustable parties without the interference of third party (Caro et al., 2018).

It creates a huge impact on the agriculture and food supply chain digital logistic network and provides the access to transfer information. Further, Agri Block IoT has been suggested as a solution to agri-food supply chain management as well as the comparison of two different blockchains and their implementation. The purpose of the introduction is to monitor the various stages in supply chain customer-oriented information to manufacturer, supplier and service provider. Digital technologies help in maintaining the relationship between innovative technology and partnership performance. In terms of literature, digital technology has been a point of significant attention. It occupies an important place in consideration of social issues and enhancing the growth of social well-being (Annosi et al., 2021; Aryal et al., 2018; Nasiri et al., 2020a, b). In consideration of the agri-food industry, digital technology and sustainable agri-food supply chain also plays an important contribution to resolve various issues such as overuse of resources. Moreover, companies try to attract more customers with the help of upgraded digital technologies. The introduction of wearable devices and gadgets helped in analysing the supply chain by recognizing the possible ways to manage and enhance the agri-food supply chain process more effectively. Logistical benefits and supervision can be achieved through using this technology (Saetta & Caldarelli, 2020; Shafique et al., 2019).

The task of producing good and adequate quality products along with maintaining standards of customer affordability has come up with a need to introduce technologies and innovative approach for increasing production without troubling mother earth. The adaptability of the sustainable agri-food supply chain is being carried out across the country and has various factors and came up with the most appropriate issues covering up the management of sustainable agri-food supply chains (Joshi et al., 2020; Lioutas & Charatsari, 2020). The development of new technologies arising from the ongoing challenges is majorly taking place with the help of the basic concept of S^3, i.e. 'sensing, smart and sustainable' (Miranda et al., 2019). It is important to present the application of S^3 technology in the field of agri-food sector for monitoring and observing field.

The agri-food supply chain is a complex industry, challenging various processes, operations and roles around the world. In addition, it does not work well with the growing number of demands and barriers imposed on it, making the need for new agri-food supply chain solutions very important. Agri-food supply chain-related stakeholders such as producers and retailers, as well as governments and police departments, are closely linked to important global challenges against the enforcement of sustainable solutions. However, as in all industries, technology plays a prime role in the implementation and decision-making process (Panetto et al., 2020).

With the advancement of the Internet of Things (IoT), smart technologies and smart devices are rapidly gaining ground in all aspects of human life. Therefore, IoT is considered an essential tool for building a strong agri-food supply chain. By reviewing the IoT and blockchain-based agri-food chain and identifying the building blocks of this type of system, we focus on key components of the IoT and blockchain, as the former lays the foundation for data collection and transmission, while this opens the way for data transparency and integrity and, more importantly, reduces the financial costs of all supply chain participants. Furthermore, the importance should not just be on a single technology but a series of technology acclaimed as 'technology stack'. This stack must be a combination of various technologies like sensing, blockchain and artificial intelligence. Moreover, blockchain and artificial intelligence can contribute to improving the agri-food chain (Hou et al., 2021; Renda, 2019).

Globalization has made food supply chains more complicated with less transparency (Jayaprakash, 2020). Centralized factories shipping their products to different parts of the country or world lead to the addition of preservatives to food products to maintain their flavour and appearance. These graded food products have diminished desirability to consumers. 3D food printing can address these challenges making the existing supply chain more flexible with increasing the value of food products as per demands of the consumer, reduction of wastage of food and more customization of automated food products thus making it more sustainable (Derrosi, 2019; Dankar, 2018).

Today traceability in food supply chains is a massive concern since humans now reside in a world that is global as well as complex and where consumers expect a high level of quality. Blockchain has been encouraged to improve traceability by providing trust. However, the practice seemed to be even more stubborn. Various studies highlight the potential of wireless and IoT nerves in the agriculture sector. When we look at the benefits of these wireless technologies, the benefits are more than enough reason to incorporate them into the supply chain sector. Machine learning is another technology that plays a similar role in achieving the goals and providing direct benefits to not just huge corporations but also small-scale industries (Dutta & Mitra, 2021).

5 Methodology

This study is an empirical and analytical research towards the identification and investigation of the factors resulting in the hesitation of an organization while introducing smart technologies in their supply chain. For this study, data has been collected by surveying few organizations in which the supply chain has a major impact. Initially, through an online meeting, a team of four subject experts were interviewed to identify the major challenges perceived by them. Further, the primary data was collected through a structured questionnaire from the employees of organizations. The questionnaire had clubbed 20 items related to employees' opinion towards the introduction of smart technologies in the supply chain. These 20 items were extracted from the challenges mentioned by the experts. A 5-point Likert scale ranging from most important (1) to least important (5) was used to assign the value of each component. A total of 21 respondents were examined or discussed with. Factor analysis was carried out to examine the collected data in SPSS 25.0 version. Factor analysis identified and grouped major factors restricting the organizations to adopt smart technologies in the supply chain. For the present study, Bartlett's test of sphericity was highly significant ($p < 0.000$), thus confirming the correlation between the attributes of the population. The communality value for each attribute came out as 0.55 or higher and has been represented in Table 1 (annexure). Total four factors were extracted as shown in Table 2 (annexure) which explains a total of 83.4% of the variance, a graph of which has also been plotted and shown in Fig. 1 (annexure). All the extracted factors are having eigenvalues of more than 1. The remaining factors were not found significant. Table 3 shows the rotated component matrix to reduce the factors on which attributes under study have high loading.

Four extracted factors based on loading of different components that account for 83% of variance are shown in Table 4 (annexure). A detailed discussion regarding the same is presented in the next section.

6 Results and Discussion

The industry participants were interviewed regarding the challenges and benefits they perceive from adopting smart technologies in their supply chain. The responses received and categorized are discussed as follows:

6.1 Perceived Challenges for Adoption of Smart Technologies

There are lots of hurdles or challenges faced by supply chain manager while introducing smart technologies in a sustainable supply chain. The collected data reflects (as shown in Table 1) that when it comes to security issue 52.4% of people think that fear of data security is most important when an organization decides to introduce smart technology in their supply chain. This fear does not emerge only in stakeholders or supply chain managers' mind but also it finds space in the customer minds. Then the fear of data misuse and passing of confidential data to the competitor comprises 33.3 and 42.9%, respectively, the next paramount barrier regarding the introduction of smart technologies in the supply chain. When we proceed further towards the technological reform, 61.9% of people thinks that the need for a proper network system for information exchange is a significant challenge to introduce smart technologies in the supply chain. Factors such as lack of trust in the new technology, lack of skilled manpower, resistance from the worker, lack of interest of employees, robots and technologies requiring more rigorous monitoring, probability of mishappening and need to increase the strength of worker which are 33.3, 28.6, 28.6, 28.6, 38.1, 28.6 and 38.1%, respectively, are the other chief factors for the introduction of smart technologies in the supply chain. Many at times people working with a limited number of technologies get so used to them that shifting to something new is rather quite difficult, often leading to mistakes. An organization needs to recruit more and more skilful worker to convert human-understandable instructions into machine language and then utilizing them as and when required. When questions were raised about the investment required for the introduction of smart technologies, the prominent challenge for an organization comes out as a need for the long-term planning focus of the organization, and it consists of 66.7%, accounted as the next important barrier regarding the introduction of smart technologies. The other chief component is the partner's rejection, and the biggest problem is the coordination issue between the partners, and it consists of 52.4% emerging as the barrier for the introduction of smart technologies in the supply chain. To identify the significant challenges and group the data into meaningful factors, factor analysis was applied, and based on outcomes from factor analysis, the following four major factors emerged as the most significant challenges for organizations to implement smart technologies in their supply chains:

Factor 1: Security Threat

This is one of the most significant factors that come out as every organization is focused to keep their information private so that their uniqueness won't get hampered. All the members in an organization and even consumers are all focused on keeping the data and maintain their privacy. If data is not secured, then the chance of breaching the data from the organization will lead the competitor to gain competitive advantages. Thus, with the introduction of smart technologies, an organization must acquire a proper network system to exchange information between all participants. One major concern raised is regarding the loyalty of the data handler. An organization must provide data handling authority to the only person who is highly trustworthy.

Factor 2: Technology Upgradation

In every organization, people are habituated to work on existing technology, but when there is a situation that occurs for the requirement of upgradation of technology, they hesitate to work on that because of not having prior experience, knowledge and sometimes because of drudgery also. Organization faces challenges while upgrading the technology because of a lack of trust in the technology. Since it is a common belief that technology favours only when it is used efficiently otherwise it brings burden. Technology upgradation seeks skilled manpower to operate so the skill of labour is also important. In general, most of the organization faces resistance from the worker before upgrading because of worker's lack of interest and physical and mental fixation on the same type of technology to which they are habituated. They don't show interest in gaining knowledge and are even afraid of a miss-happening in the workplace because of poor coordination between worker and robots.

Factor 3: High Investment

It is the most significant factor that comes before installation and describes how much investment is required to introduce smart technologies. It is assumed that smart technologies seek more investment which brings hesitation among the stakeholder to finance such a huge amount of money. This high investment is the biggest barrier to the introduction of smart technologies in the supply chain. The introduction of smart technology not just leads to high investment but also less profit in the initial stage depending on the type of organization. Even if there is the slightest chance of failure, it brings ambiguity of decision in the mind of stakeholder and makes them resistant to invest huge amount of money.

Factor 4: Partner's Rejection

When the decision to the introduction of smart technologies comes in front of the partner of the supply chain whose profit is based on an incentive basis, they hesitate to introduce smart technology. The advantages of smart technology like transparency and traceability strengthen the relationship between partners, but at some point, it also affects the business process of different partners of the supply chain. The advancement of technology makes the supply chain responsible, reduces lead time and improves coordination; however, sometimes these benefits are overlooked before personal gains.

6.2 Perceived Benefits from Smart Technologies

Also, from the primary data collection, it emerged that with the introduction of smart technologies, the performance of the supply chain increases up to a great extent which reduces the drudgery of labour, enhances the efficiency of labour, makes them time efficient, etc. Smart technologies are beneficial for supporting the various prospects of any organization. A few of the prospects perceived by industry participants are listed below:

6.2.1 Cost-Efficient

The use of smart technologies helps in the reduction of wastages of resources, makes things efficient and reduces labour, thus helping in the deduction of manufacturing cost which leads to selling product at a low cost; therefore, the company can go for a cost-efficient strategy.

6.2.2 Lead Time Reduction

Smart technologies help in maintaining the proper coordination between the partners by providing the access to trace the transportation system and track the outbound logistics; thus the lead time gets reduced.

6.2.3 Decrease Inventory Loss

Sharing the proper information and proper anticipation of upcoming demand by tracking previous order and providing the proper ambience to store smart technology helps in the reduction of inventory loss.

6.2.4 Responsive

Smart technologies help in the reduction of manufacturing time, lead time, and allow the organization to fulfil customer demand in the desired situation. These technologies deduce inventory loss and thus help to be responsive.

6.2.5 Reduce Uncertainty

Smart technologies help in anticipation of incoming order and maintaining the track of outbound logistics. By this, an organization can adjust its inbound and outbound logistics, thus gaining more and more surplus.

6.2.6 Proper Coordination

Smart technologies aid in the proper coordination of the supply chain by bringing in more transparency and abolishing information delays. Thus, organizations can make better coordination of material, statistics and finances.

6.2.7 Reduce Bullwhip Effect

The incorporation of smart technologies into the supply chain empowers an organization by leading the way towards accurate demand forecast. This allows the organization to make better resolutions of inventory requirement, thus, reducing the bullwhip effect.

6.2.8 Increase in Whole Supply Chain Surplus

The utmost objective of any supply chain is to bring maximal supply chain surplus. Smart technologies assist in achieving this goal by not just managing inventory but also customer relationship and other operational expenses.

6.2.9 Competitive Advantage

Smart technologies bring innovation into the supply chain. Like, machine learning, artificial intelligence and 3D printing bring automation into the supply chain which ultimately leads to delivering competitive advantages.

6.2.10 Creating Value for the Consumer

Supply chains with smart technologies bring forth hastened product delivery at a better price for the customer. Further, by making customized products organizations can polish their relationship with their customers by creating value for them.

6.2.11   Providing the Longevity

When a supply chain works on smart technologies, its functioning is based on real-time data. This allows an organization to maximize their ROI and minimize its losses. Eventually, it removes physical barriers and brings longevity.

6.2.12   Protecting the Environment

Smart technologies help to make essential changes in the supply chain to make it environmentally sustainable. Pursuits, like ethical sourcing and reduction of fossil fuel consumption, help in the protection of the environment.

6.2.13   Reducing Waste

The supply chain with smart technologies like machine learning speeds up delivery, enhances the quality and reduces waste. It also helps in the reduction of waste during the product flow among components of supply chains.

6.2.14   Society Upliftment

Smart technologies help in the upliftment of society by improving the working condition of employees. Smart technologies make employees skilful and provide them with proper wages. This makes organizations, as well as society, empowered.

6.3   Strategies for Implementation of Sustainable Agri-Food Supply Chain

For an organization, the most important thing is the satisfaction of their customer, which makes the customer loyal to the organization. This loyalty can be gained by providing value to the customer by the means of making a cost-efficient supply chain or responsive supply chain. Today, consumers and society are not only fixed to their benefits but are also focused on environmental protection and social upliftment. For this, organizations are forced to introduce smart technologies in their supply chain. These technologies help to protect the environment, enhance the working condition of the worker, provide value to the consumer, increase longevity to their respected supply chain, provide competitor advantages and ultimately generate large revenue. Thus, the sustainable supply chain needs to have a strong relationship with its partners. Moreover, these collaborations between shareholder and supply chain manager are important, since they both are actively involved in taking concern about the environment and longevity of the supply chain. In any type of organization, these smart technologies cannot be introduced suddenly so there are some strategies to bring these technologies in the action. These strategies are discussed below:

6.3.1 Convincing Stakeholders

For introducing new technologies large investment is required along with patience. So, the prominent function of a supply chain manager is to convince the stakeholder to invest money. This action is required because either the stakeholders are not interested to invest or they can't wait for some time to let the technologies get involved and generate desired output in the supply chain.

6.3.2 Recruiting More Technical Staff

These technologies work on a programming language so an organization has to appoint more and more technical staff so that they can convert human instruction into machine-understandable language. Any wrong instruction to a machine can cause serious issues in the working environment; hence, skilful staff must be there. These staff must also help in maintaining the collaboration between human and robot. The harmonious coordination between the skilful worker and supply chain manager leads to the enhancement in the performance of the supply chain.

6.3.3 Provide Training to Existing Staff

New technologies have new functions and new method of operations. To make it work, new skills are also required, but usually, the existing staff who are habituated to working on their old skills are reluctant towards upgradation of skills. An organization must train them on the latest technologies and polish their skills so that they can work with more efficiency and ease. These new technologies make them multifunctional and efficient and provide access to agile working.

6.3.4 Profit Distribution Based on Revenue Generation

The organization used to provide profit to some of their partners or sometimes the third party in the form of incentive. This makes them use different means to generate profit for themselves without adversely affecting the whole supply chain. But with the introduction of smart technologies, their adulteration can be caught so they don't support including them in the organization. Thus, to involve them and let them function properly, an organization must give profit from the whole supply chain surplus instead of based on incentive. This profit share brings loyalty to them which could enhance the performance and add value to the supply chain.

6.3.5 Focus to Secure the Data

The introduction of smart technologies brings transparency to the system. To provide transparency the data should be connected end to end and partners must have access to monitor that. This end-to-end connection may cause the leaking of data either by the competitor or by unfair means of their partners themselves. To get rid of these types of fraudulence, an organization must make their security function so efficiently that their data cannot be misused. This step also brings confidence to their customers that their data is safe and will not be misused. This brings value to them and increases their loyalty.

6.3.6 Government Approval

Supply chain manager should convince the government to let them bring smart technologies at less tax rate, if possible, provide subsidies and help the organization to borrow money from the bank or any other government sector. This could be achieved by convincing government officers on how these technologies are dealing with the various environmental concerns and their role in social upliftment.

7 Conclusion

The supply chain is most important to the organization. It consists of the actions of all the partners, and most importantly it connects the consumer which is the source of income for the partners. So, the supply chain should be sustainable and must provide value to the consumer. The involvement of smart technologies helps in the removal of the drudgery of labour, making it environmentally safe, reducing wastage and uplifting working condition. Moreover, it helps in uplifting society, gaining competitive advantages, enhancing the life of the supply chain, providing transparency and providing access to work from any place. It also provides values and makes the supply chain cost-efficient as well as responsive. These technologies not only favour the manufacturer but also provide benefits to all the members connected to that organization.

Although there are a lot of hurdles in introducing the smart technologies in the agri-food supply chain, where investment is amongst the top contributor, still companies are adopting digital technologies to gain competitive advantage. It is evident that these technologies may not provide enormous benefit at an instant but with the passing time, it improves the organization status. In the current scenario where there is lots of competition from both local and global companies, an organization must improve itself by taking technological advantages and showing concern towards the consumer and environment.

Annexure

Table 1 Communalities

Sl. no	Communalities	Initial	Extraction
1.	Need proper network system for information exchange	1.000	0.826
2.	Fear of data security	1.000	0.725
3.	Chances of data misuse	1.000	0.788
4.	Lack of trust in the new technology	1.000	0.816
5.	Lack of skilled manpower	1.000	0.751
6.	Resistance from workers	1.000	0.911
7.	Lack of interest of employees	1.000	0.879
8.	Need extensive training for the employees	1.000	0.917
9.	Need for long-term planning focus of the organization	1.000	0.650
10.	Need for relating technology adoption with incentives	1.000	0.854
11.	Robots and technologies require more rigorous monitoring	1.000	0.879
12.	Probability of mishappening	1.000	0.922
13.	Need to increase the strength of workers	1.000	0.833
14.	Chances of confidential information going to competitors	1.000	0.841
15.	A large number of investments are required	1.000	0.779
16.	Shareholders might not be interested	1.000	0.790
17.	The initial implementation period will be disturbing	1.000	0.891
18.	Coordination issue between the partners	1.000	0.869
19.	Profit distribution (even distribution vs. incentive-based)	1.000	0.897
20.	Transparency in the system may not be welcomed by all partners	1.000	0.862
	Extraction method: Principal component analysis		

Table 2 Total variance explained (eigenvalues)

Total variance explained

Component	Initial eigenvalues			Extraction sums of squared loadings			Rotation sums of squared loadings		
	Total	% of Variance	Cumulative (%)	Total	% of Variance	Cumulative (%)	Total	% of Variance	Cumulative (%)
1	10.741	53.706	53.706	10.741	53.706	53.706	5.719	28.597	28.597
2	2.684	13.421	67.128	2.684	13.421	67.128	4.603	23.015	51.612
3	2.103	10.513	77.641	2.103	10.513	77.641	3.944	19.720	71.332
4	1.152	5.761	83.402	1.152	5.761	83.402	2.414	12.070	83.402
5	0.840	4.202	87.604						
6	0.647	3.234	90.838						
7	0.463	2.313	93.152						
8	0.411	2.057	95.208						
9	0.304	1.520	96.729						
10	0.247	1.234	97.963						
13	0.072	0.359	99.668						
14	0.043	0.213	99.881						
15	0.015	0.073	99.954						
16	0.008	0.039	99.993						
17	0.001	0.007	100.000						
18	5.318E-16	2.659E-15	100.000						
19	4.475E-16	2.238E-15	100.000						

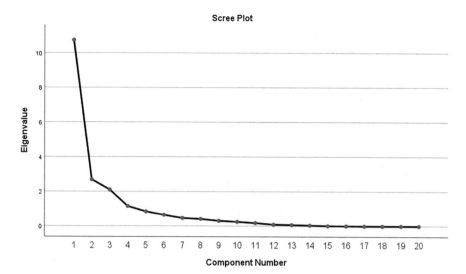

Fig. 1 Screen plot (eigenvalue)

Table 3 Rotated component matrix

Sl. no.	Challenges	Component			
		1	2	3	4
1.	Need proper network system for information exchange	−0.110	0.629	0.120	0.636
2.	Fear of data security	0.515	0.610	0.288	0.060
3.	Chances of data misuse	0.278	0.528	0.097	0.650
4.	Lack of trust in the new technology	0.752	0.315	0.375	−0.107
5.	Lack of skilled manpower	0.585	0.281	0.572	−0.053
6.	Resistance from workers	0.806	0.198	0.467	0.064
7.	Lack of interest of employees	0.890	0.102	0.270	−0.052
8.	Need extensive trainings for the employees	0.321	0.855	0.284	0.047
9.	Need for long-term planning focus of the organization	0.201	0.765	0.028	0.152
10.	Need for relating technology adoption with incentives	0.301	0.846	0.137	0.172
11.	Robots and technologies require more rigorous monitoring	0.659	0.649	0.014	0.152
12.	Probability of mishappening	0.528	0.030	0.425	0.679
13.	Need to increase the strength of workers	0.828	0.371	0.065	0.078
14.	Chances of confidential information going to competitors	0.838	0.277	0.111	0.223
15.	Large number of investments are required	0.127	0.070	0.847	0.201
16.	Shareholders might not be interested	0.367	0.185	0.755	0.227
17.	Initial implementation period will be disturbing	−0.143	0.070	0.523	0.770
18.	Coordination issue between the partners	0.622	0.506	0.359	0.314
19.	Profit distribution (even distribution vs. incentive-based	0.344	0.590	0.639	−0.148
20.	Transparency in system may not be welcomed by all partners	0.066	0.105	0.827	0.405
	Extraction method: Principal component analysis rotation method: Varimax with Kaiser normalization				
	a. Rotation converged in 19 iterations				

Table 4 Factors extracted from factor analysis

Sl. no.	Challenges	
	Factor description	Variables included
1.	*Security threat*	
a.		Fear of data security
b.		Chances of data misuse
c.		Chances of confidential information going to competitors
2.	*Technology upgradation*	
a.		Need proper network system for information exchange
b.		Lack of trust in the new technology
c.		Lack of skilled manpower
d.		Resistance from workers
e.		Lack of interest of employees
f.		Robots and technologies require more rigorous monitoring
g.		Probability of mishappenings
h.		Need to increase the strength of workers
3.	*High investment*	
a.		Need extensive training for the employees
b.		Need for long-term planning focus of the organization
c.		Need extensive training for the employees
4.	*Partner's rejection*	
a.		Transparency in the system may not be welcomed by all partners
b.		Coordination issue between the partners
c.		Shareholders might not be interested
d.		Profit distribution (even distribution vs. incentive-based)

Acknowledgment The authors are thankful to the National Institute of Food Technology Entrepreneurship and Management (NIFTEM), Kundli, Sonepat (Haryana), India, and the Ministry of Food Processing Industries (Govt. of India), for providing essential support and funding to conduct this research. Grant Number: Q-11/13/2021-R&D.

References

Abbasi, M., & Nilsson, F. (2012). Themes and challenges in making supply chains environmentally sustainable. *Emerald Insight, 17*(5). https://doi.org/10.1108/13598541211258582

Annosi, M. C., Brunetta, F., Bimbo, F., & Kostoula, M. (2021). Digitalization within food supply chains to prevent food waste. Drivers, barriers and collaboration practices. *Science Direct Journal of Industrial Marketing Management, 93*, 208–220. https://doi.org/10.1016/j.indmarman.2021.01.005

Aryal, A., Nattuthurai, P., Ying, L., & Bo, L. (2018). The emerging big data analytics and IoT in supply chain management: A systematic review. *Supply Chain Management, 25*(2), 141–156. https://doi.org/10.1108/SCM-03-2018-0149

Asaf, T. (2020). *Artificial intelligence for agricultural supply chain risk management: Constraints and potentials.* CGIAR. https://hdl.handle.net/10568/108709

Ben-Daya, M., Hassini, E., & Bahroun, Z. (2019). Internet of things and supply chain management: A literature review. *International Journal of Production Research, 57*(15–16), 4719–4742. https://doi.org/10.1080/00207543.2017.1402140

Caro, M. P., Ali, M. S., Vecchio, M., & Giaffreda, R. (2018). Blockchain-based traceability in Agri-Food supply chain management: A practical implementation. *2018 IoT Vertical and Topical Summit on Agriculture Tuscany.* http://ieeexplore.ieee.org/stamp/stamp.jsp?tp=&arnumber=8373021&isnumber=8373016.

Chopra, S., & Meindl, P. (2016). *Supply chain management: Strategy, planning, and operation* (6th ed.). Pearson.

Dankar, I., Haddarah, A., Omar, F. E. L., Sepulcre, F., & Pujolà, M. (2018). 3D printing technology: The new era for food customization and elaboration. *Trends in Food Science & Technology, 75*, 231–242. https://doi.org/10.1016/j.tifs.2018.03.018

Derossi, A., Husain, A., Caporizzi, R., & Severini, C. (2019). Manufacturing personalized food for people uniqueness. An overview of traditional to emerging technologies. *Critical Reviews in Food Science and Nutrition, 60*, 1–19.

Duong, L. N. K., Al-Fadhli, M., et al. (2020). A review of robotics and autonomous systems in the food industry: From the supply chains perspective. *Journal of Trends of Food Science & Technology, 106*, 355–364. https://doi.org/10.1016/j.tifs.2020.10.028

Dutta, P. K., & Mitra, S. (2021). *Application of agricultural drones and iot to understand food supply chain during post COVID*-19. Agricultural informatics: Automation using the IoT and machine learning, Chapter 4. Wiley Online Library. https://doi.org/10.1002/9781119769231.ch4.

Dyllick, T., & Hockerts, K. (2002). Beyond the business case for corporate sustainability. *Business Strategy and the Environment, 11*(2), 130–141.

Hou, L., Liao, R., & Luo, Q. (2021). IoT and blockchain-based smart agri-food supply chains. In J. C. Augusto (Ed.), *Handbook of smart cities.* Springer. https://doi.org/10.1007/978-3-030-15145-4_91-1.

Jayaprakash, S., Paasi, J., Pennanen, K., Ituarte, I. F., Lille, M., Partanen, J., & Sozer, N. (2020). Techno-economic prospects and desirability of 3D food printing: Perspectives of industrial experts, researchers and consumers. *Food, 9*, 1725. https://doi.org/10.3390/foods9121725

Joshi, S., Singh, R. K., & Sharma, M. (2020). Sustainable Agrifood supply chain practices: Few empirical evidences from a developing economy. *SAGE Journals, 10*, 1177.

Lioutas, E. D., & Charatsari, C. (2020). Smart farming and short food supply chains: Are they compatible? *Journal of Land Use Policy, 94*, 104541. https://doi.org/10.1016/j.landusepol.2020.104541

Miranda, J., Ponce, P., Molina, A., et al. (2019). Sensing, smart and sustainable technologies for Agri-food 4.0. *Computers in Industry, 108*, 21–36. https://doi.org/10.1016/j.compind.2019.02.002

Mostafa, N., Hamdy, W., & Alawady, H. (2019). Impacts of internet of things on supply chains: A framework for warehousing. *Journal Social Sciences, 2019*(8), 84. https://doi.org/10.3390/socsci8030084

Nasiri, M., Ukko, J., Saunila, M., & Rantala, T. (2020a). Managing the digital supply chain: The role of smart technologies. *Journal of Technovation, 9697*, 102121. https://doi.org/10.1016/j.technovation.2020.102121

Nasiri, M., Ukko, J., Saunila, M., et al. (2020b). Managing the digital supply chain: The role of smart technologies. *Elsevier Technovation, 96–97*, 102121.

Panetto, H.,Lezoche, M.,Hernandez, J., Diaz, M. M. E. A., & Kacprzyk, J. (2020). Special issue on Agri-Food 4.0 and digitalization in agriculture supply chains – New directions, challenges and applications. *Computers in Industry*, Elsevier. DOI https://doi.org/10.1016/j.compind.2020.103188.

Quinnell, R. (2019). *How sensors and networks are empowering 'Agriculture technology'.* EPS News.

Renda, A. (2019). The age of Foodtech: Optimizing the Agri-food chain with digital technologies. In R. Valentini, J. Sievenpiper, M. Antonelli, & K. Dembska (Eds.), *Achieving the sustainable development goals through sustainable food systems*. Springer. https://doi.org/10.1007/978-3-030-23969-5_10

Saetta, S., & Caldarelli, V. (2020). How to increase the sustainability of the Agri-food supply chain through innovations in 4.0 perspective: A first case study analysis. *Procedia Manufacturing, 42*, 333–336. https://doi.org/10.1016/j.promfg.2020.02.083

Santiteerakul, S., Sopadang, A., et al. (2020). The role of smart Technology in Sustainable Agriculture: A case study of Wangree plant factory. *Sustainability, 2020*(12), 4640.

Sardar, S. K., Sarkar, B., & Kim, B. (2021). Integrating machine learning, radio frequency identification, and consignment policy for reducing unreliability in smart supply chain management. *MDPI Journals, 9*(2), 247. https://doi.org/10.3390/pr9020247

Saunila, M., Nasiri, M., Ukko, J., & Rantala, T. (2019). Smart technologies and corporate sustainability: The mediation effect of corporate sustainability strategy. *Journal of Computers in Industry, 108*, 178–185. https://doi.org/10.1016/j.compind.2019.03.003

Saurabh, S., & Dey, K. (2020). Blockchain technology adoption, architecture, and sustainable agri-food supply chains. *Journal of Cleaner Production., 284*, 124731. https://doi.org/10.1016/j.jclepro.2020.124731

Seuring, S. (2012). A review of modelling approaches for the sustainable supply chain management. *Elsevier Journal Decision Support System, 54*(4), 1513–1520.

Shafique, M. N., Khurshid, M. M., Rahman, H., Khanna, A., Gupta, D., & Rodrigues, J. J. P. C. (2019). The role of wearable Technologies in Supply Chain Collaboration: A case of pharmaceutical industry. *IEEE Access, 7*(2019), 49014–49026.

Voon, S. L., An, J., Wong, G., Zhang, Y., & Chua, C. K. (2019). 3D food printing: A categorised review of inks and their development, virtual and physical prototyping. *Science Direct, 4*(3), 203–218. https://doi.org/10.1080/17452759.2019.1603508

Yang, F., Zhang, M., & Bhandari, B. (2017). Recent development in 3D food printing. *Critical Reviews in Food Science and Nutrition., 57*(14), 3145–3153. https://doi.org/10.1080/10408398.2015.1094732

Wireless Sensors' Location for Smart Transportation in the Context of Industry 4.0

Mustapha Oudani, Sarah El Hamdi, Abderaouf Benghalia, Imad El Harraki, Hanane El Raoui, and Karim Zkik

1 Introduction

Transportation is the cornerstone of each logistics chain. The authors in (Bubelíny et al., 2021) identify in the literature that 1.1 billion cars and 400 million trucks were on the world's roads in 2015, and in 2040 there will be 2 billion cars and 800 million trucks. There are two types of transport, the transport of people (travelers) and the transport of goods. They are considered two independent systems, although they may use the same infrastructure but rarely the same vehicles. Transportation costs have a huge impact on the prices of goods. Indeed, with the strong competition on

M. Oudani (✉)
TICLab, ESIN, International University of Rabat, Rabat, Morocco
e-mail: mustapha.oudani@uir.ac.ma

S. El Hamdi
Laboratory of Engineering Science, MOSIL Team, ENSA, Kénitra, Morocco
e-mail: sarah.elhamdi@uir.ac.ma

A. Benghalia
Algiers I University, Algiers, Algeria

I. El Harraki
Industrial Engineering Department, National Superior School of Mines, Rabat, Morocco
e-mail: elharraki@enim.ac.ma

H. El Raoui
Department of Comp. Science and A.I., Universidad de Granada, Granada, Spain

Modeling and Mathematical structures Laboratory, Sidi Mohamed Ben Abdellah University, Fez, Morocco
e-mail: hanane.elraoui@usmba.ac.ma

K. Zkik
ESAIP Graduate School of Engineering, CERADE, Angers, France

© The Author(s), under exclusive license to Springer Nature Switzerland AG 2023
S. S. Kamble et al. (eds.), *Digital Transformation and Industry 4.0 for Sustainable Supply Chain Performance*, EAI/Springer Innovations in Communication and Computing, https://doi.org/10.1007/978-3-031-19711-6_4

the markets, the main challenge is to produce and deliver as quickly as possible by reducing the various costs and by satisfying the customers who are generally geographically separated and are located further and further from the centers' delivery. There are several modes of transport, among others: road transport, sea or river transport, and rail transport. The road transport business is a sensitive activity that requires controlling the flow of goods and improving the performance of distribution to customers in order to ensure delivery in order to meet their requirements. It is obvious that customers need to be delivered quickly and regularly in order to keep the balance of their business. However, if by bad luck an accident does occur, then and consequently the delivery operation should be restarted from the beginning. This results in a waste of time as well as high costs and all lead to poor traffic by complicating the tasks of the carriers. Intelligent transport then becomes an essential component of modern societies (Karami & Kashef, 2020). In addition, urban traffic is at the heart of many issues and has become an essential aspect of daily life. The latter has grown in the space of a few years, causing many problems that cost time, money, health, and environmental quality every day, whether through traffic jams, accidents, or even offenses. A potential alternative for road transportation is the intermodal transportation (Oudani, 2021). As part of the fourth industrial revolution (Industry 4.0), the trend of logistics is to an intelligent logistics system (Bag et al., 2020). In recent years, interconnection and interoperability in supply chains is often used as a pertinent solution. The objective is to reduce traffic and minimize the impact of the vehicle on the environment, but not only by connecting cars that drive themselves but also by making intelligent transport and logistics systems. Industry 4.0 is evolving rapidly and can have global impacts in almost any industry. Supply chain management (SCM) is facing several challenges including agility, efficiency, and durability (El Raoui et al., 2020). An intelligent transport system consists of integrating new information and communication technologies in the field of transport. The intelligence aspect makes it possible to improve the transport system through the various functions such as traceability, communication, adaptive behavior, information processing, and learning. However, the development of smart cities requires us to acquire real-time information on road infrastructure. A smart city aims to improve the quality of urban services and offer smart mobility with a lower cost by using information and communication technologies (ICT). In these smart cities we find "intelligent transport systems" (ITS) which are advanced applications or services combining transport engineering, communication technologies, and information and geographic positioning. Indeed, the Digital Internet transparently allows interconnection between networks, thus authorizing the transmission of data packets formatted in a standard manner allowing them to transit through heterogeneous equipment complying with the TCP/IP protocol. A set of objects that works thanks to infrastructures and technologies linked by an Internet connection is called connected objects (Iot). The control of connected objects is done via a smartphone, tablet, or computer. They have become everyday objects and they are present in complex systems like vehicles, airplanes, etc. They are also present in homes: refrigerator, heating system, television, electric garage door, etc. The Physical Internet aims to exploit the metaphor of the Internet in order to propose a vision for a durable and progressively deployable solution to the global problems related to the way we transport, handle, store, realize, supply, and use physical objects across the planet. In the field of logistics, the use of sensors can improve the performance of inventory management, transport, and traceability of goods. Sensors can also be used to improve environmental performance by monitoring temperature, air quality, etc. In transportation, the Internet of Things is considered the biggest revolution in the industry. IoT impacts the entire supply chain, from sensors that track goods to wireless connectivity. Their use allows better communication between devices deployed in the field and logistics software, in order to monitor and manage flows in real time. Now with the development of technologies of wireless sensors, the main issue is to allow each network element distributed to collaborate with other connected components and can make decisions improving the performance of the transport system. The elements of this type of network generally have computational and memory capacities, allowing them to perform coordination exchanges. Wireless sensor networks are affordable because they are made up of a large set of sensors with generally limited energy and capacity (Faye & Chaudet, 2015). Transport logistics can be broken down into three activities: fleet management, elaboration of transport networks, and transport planning. The routes are determined by focusing on:

1. Transport risks (use of intermodal transport units, limiting the number of load breaks)
2. Product transporting (road, air, maritime, rail, multimodal)
3. Choosing the cheapest network and best suited to the goods to be transported
4. Logistics resources availability (handling engines, storage areas)

Managing a complex transportation system requires good planning and relevant machine learning (Karami & Kashef, 2020). The objective is to be able to predict traffic flows and forecast based on the various parameters of the past such as traffic speeds, taking into account road conditions.

2   Literature Review

Industry 4.0 is based on innovation or experimentation with ideas using new physical, digital, and biological technologies. The aim is to improve the resource efficiency and environmental performance at different levels of the supply chain. The principle of the Physical Internet (IoT) is to interconnect independent logistics networks through accessible resources, standardization of interfaces and protocols, smart containers, etc., in order to improve integration, coordination, and sharing resource. Proper organization of a transport chain for a customer requires identifying the mode and/or combination of transport modes with appropriate "transfer points" to enable overall performance. The issue addressed by (Qiao, 2018) is the Revenue Management issue for LTL carriers in the Physical Internet. The objective is the application of Revenue Management in a dynamic transport network such as the

Physical Internet, in order to provide a solution to improve the incomes of operators. The IoT makes it possible to have information concerning the road flow in real time in order to offer optimized utilities and contribute to the offer of an intelligent transport system (Buyya & Dastjerdi, 2016) which makes it possible to collect, store, and process the data set (Benevolo et al., 2016). In the same context of the evolution of intelligent transport, a set of connected trucks makes it possible to contribute to the optimization of the fleet by exploiting the automated processing of information, for example, the level of driver fatigue and the end of the working hours, to be able to suggest and reserve a parking space and help the driver to park safely (Daws, 2021). Furthermore, data collection solutions based on wireless sensor networks (WSNs) were considered from the early days of the Internet of Things (IoT) paradigm due to their important features such as sensing capabilities and long-term, low-cost, versatile actuators (Song et al., 2014). Indeed, a wireless sensor network (WSN) may be defined as a network of small sensors communicating using wireless technology for receiving and sending data (Bottero et al., 2013). These networks are a combination of wireless communication and distributed sensing using self-powered device with limited compute and memory functions. Advances in sensors have made it possible to develop intelligent transport systems based on reliable, frequent, and large-scale data. They allow measurements of the character-istics of vehicles, traffic (flow, occupancy rate, speed, etc.), or events (incidents, queues, crossing red lights, etc.) to also optimize and improve road safety. The work carried out aims to study the use of distributed systems to implement intelligent transport systems through a network of wireless sensors. According to (Peixoto & Costa, 2017), different types of devices ensuring high levels of safety are adopted by governments. These devices, such as wireless visual sensor networks (WVSN), allow cities to be monitored without the use of cables. In Peixoto and Costa (2017), an algorithm was proposed to study the problem of positioning and locating mobile points in WVSN installed on streets and roads. The proposed approach makes it possible to take into account the constraints of smart cities by detecting prohibited areas in order to position sensor points in authorized areas. In (Bottero et al., 2013), the authors have proposed a WSN for traffic control in the logistics hub area in Turin. The reference (Jiang et al., 2020) conducted a study about intelligent logistics monitoring systems based on sensor networks and big data. In this work, the objective is to study and assess the information flow of logistics operations, namely, departure and arrival, warehouse management, positioning, and monitoring of distribution. The transport of people and goods to their destination faces several problems because of the high number of traffic jams and the capacity of infrastruc-ture saturated by the increased evolution of the number of vehicles on the roads (Bubelíny et al., 2021). The new trend in cities is to become intelligent by working on mobility and the quality of life of the population. Indeed, it is a question of reducing CO_2 emissions at the level of agglomerations and urban areas and therefore takes into consideration transport which has a significant impact on the environment. A smart city must improve the quality of life of its population while minimizing its environmental impact. The paper (Benevolo et al., 2016) affirms the direct effect of intelligent mobility on the population of cities and the direct link between the quality of life and the challenges of intelligent mobility. In (Ning et al., 2014), connected vehicles are equipped with technology to increase their reliability, efficiency, and safety while improving driver comfort. In (Bubelíny et al., 2021), the authors addressed that a smart city should consider and include stations if rail mode is available. Indeed, the availability and location of stations are key elements of rail transport. In (Porru et al., 2020), authors conducted a study on the use of the Internet of Things to improve the quality of public transport (PT) services in rural and urban areas. They identified three main elements concerning rural and urban mobility, namely, the detection and dynamic definition of optimized routes as well as the simplification of the management of investment planning. In (Karami & Kashef, 2020), authors reviewed several research activities using machine learning techniques for the development of smart transportation. They presented different models and machine learning techniques using time series prediction for intelligent transportation planning.

3   Wireless Sensor Networks for Transportation

Wireless sensor networks (WSNs) refer to a very large number of spatially dispersed electronic devices called sensors. The main functionalities of sensors are (i) sensing, (ii) communication, and (iii) data processing. This type of networks has several applications in various domains like waste and water monitoring, natural disaster prevention, air pollution monitoring, landslide detection, and so on. A typical example of sensor networks are those used to measure environmental conditions such as humidity, sound, temperature, sound, and wind.

Most applications from Industry 4.0 using Internet of Things are based on wireless sensor networks. Nevertheless, several authors have pointed out several issues that may hinder the use of WSNs (Souissi et al., 2019; Mohamed et al., 2017). In fact, in most cases, sensors are tiny devices with small computing capacity, small memory and connectivity, and rudimentary energy resourcing. Several papers in literature have studied WSN issues. We can cite security challenges (Patil & Chen, 2013), performance assessment (Dima et al., 2014; Sivakumar & Radhika, 2018), quality of services (Hashish et al., 2017), and energy efficiency (Thangaramya et al., 2019; Tao et al., 2015).

Energy efficiency represents one of the most important issues addressed in the literature as it is linked to many parameters such as the increasing number of sensor devices and their multiple and heterogeneous energy resourcing forms and their non-traditional location of devices that make their charging very difficult (e.g., cars, bodies, forests, homes). To address energy efficiency issues, there are multiple parameters that must be taken into consideration such as communication between sensors, routing, sink location, and coverage.

Recently, energy efficiency problems have attracted the attention of many researchers. These problems were addressed using various methodologies and different solving and modeling approaches (Sivakumar & Radhika, 2018; Rafael Asorey-Cacheda et al., 2017). The authors in (Touati et al., 2017) give an overview on optimization techniques for energy consumption in WSNs. They present a classification of the reviewed papers into three categories including data structure, operating time partition, and sensor mobility. The authors of (Elhoseny et al., 2018) studied the K-coverage problem of mobile sensors in WSNs. They solved the problem using genetic algorithm. The authors in (Güney et al., 2012) studied the energy minimization on WSNs. They formulated this problem using two modeling approaches: the first based on $p-$median and the second one on the single-commodity network flow. Nevertheless, this paper studied a specific case where the potential sites to locate sinks and their number are known in advance and without considering the dynamic aspect of the network. Accordingly, this restrictive assumption is being a constraint for applying the model in all types of sensor networks.

We will study in this chapter the wireless sensor placement problem using fuzzy linear programming approach as well as observability approach. In the first step, we present fuzzy mixed integer models minimizing energy consumption. A fuzzy constraint for a "given level of coverage" is added to model the decision-maker to locate sensors. Fuzzy budget constraint will be formulated using a fuzzy inequality. In the second step, we study the observability of a traffic network for the placement of wireless sensors.

3.1   Single-Commodity Network Flow

The single-commodity network flow is a strategic problem aiming to design the WSN. The objective is to minimize the total energy (or the total cost of the energy consumption) consumed by the sensors within the network (Oudani & Zkik, 2019). Specifically, the energy of the batteries is used for transmitting data flows between sensors and sinks. To formulate the problem, we should consider the two following decisions: (i) a placement component deciding on the location of sensors and sinks and (ii) a routing component deciding on the transmission of data packets between sinks and sensors. Solving these two independent decisions may contribute in the extension of sensors and sink lifetimes. Reference (Güney et al., 2012) provides a single-commodity network flow formulation for wireless sensor network.

The parameters used for the problem modeling are the following:

I : the field of sensors $|\ I\ | = N$: the number of potential points to locate sensors within the field
$\mathcal{T}$: the set representing all types of sensors $|\ \mathcal{T}\ | = K$: the number of sensor types
q_{ijk}: the coverage capacity of sensor type-k at point j located at point i
θ_j: the coverage threshold at point j
β_{ij}: the distance between i and j
h_k: the cost of deploying a sensor of type-k

e_{ijk}: the energy consumed per unit data flow from sensors to sinks and from sensors to sensors

B: the limit budget to deploy sensors

p: the number of sinks to be added

The problem has the following decision variables:

μ_{ijkl}: the quantity of data flow from a sensor type k located at point i to a sensor type l located at node j

v_{ijk}: the quantity of data flow from a sensor type k located at point i to a sink located at node j, with $j \neq i$

s_{ik}: a 0–1 decision variable taking value 1 if the sensor type-k is deployed at point i and 0 otherwise

z_j: a 0–1 decision variable, equal 1 if a sink is located at point j and 0 otherwise

w_i: represent the total data inflow to a point i when a sink is located at that point.

The Coverage Sink Location and Routing Problem (CSLRP) can be formulated as a single-commodity network flow and written as follows:

$$\text{Min} \sum_{i \in I} \sum_{k \in T} \sum_{j \in I} \sum_{l \in T} e_{ijk} \mu_{ijkl} + \sum_{i \in I} \sum_{k \in T} \sum_{j \in I} e_{ijk} v_{ijk} \tag{1}$$

The linear objective expressed by the Eq. (1) aims to minimize the total consumed energy between sensors and sinks and between a sink and other sinks. This objective function is subject to several constraints such as:

$$\sum_{i \in I} \sum_{k \in T} q_{ijk} s_{ik} \geq \theta_j, \forall j \in I \tag{2}$$

The constraint (2) guarantees that the coverage threshold is satisfied in every node of the field.

$$\sum_{i \in I} \sum_{k \in T} h_k s_{ik} \leq B \tag{3}$$

The constraint (3) guarantees that the overall cost of sensor deployment does not exceed the allocated budget.

$$\sum_{j \in I} \sum_{l \in T} \sum_{k \in T} u_{jilk} + \sum_{j \in I} \sum_{l \in T} v_{jil} + \sum_{k \in T} = \sum_{j \in I} \sum_{l \in T} \sum_{k \in T} u_{ijlk} + \sum_{j \in I} \sum_{k \in T} v_{ijk} + w_i, \forall i \in I \tag{4}$$

The constraint (4) expresses the fact that the total inbound flows to a node of the field equal the total outbound flows from that node.

$$\sum_{j \in I} \sum_{l \in \mathcal{T}} \mu_{ijkl} + \sum_{j \in N} v_{ijl} \leq NKs_{ik}, \forall i \in I, k \in \mathcal{T} \tag{5}$$

The constraint (5) guarantees sending the data flow through the deployed sensors.

$$\sum_{i \in I} \sum_{k \in \mathcal{T}} v_{ijk} \leq NKz_j, \forall j \in I \tag{6}$$

The constraint (6) ensures the transmission of data flow from sensors to a sink if the later is deployed.

$$w_i \leq NKy_i, \forall i \in I \tag{7}$$

The constraint (7) is used to guarantee that data is absorbed only by sinks.

$$\sum_{j \in N} z_j = p \tag{8}$$

The constraint (8) guarantees locating exactly p sinks.

$$\mu_{ijkl} \geq 0, \forall i, j \in I, k, l \in \mathcal{T} \tag{9}$$

$$v_{ijk} \geq 0, \forall i, j \in I, k \in \mathcal{T} \tag{10}$$

$$w_i \geq 0, \forall i \in I \tag{11}$$

$$y_i, s_{ik} \in \{0, 1\} \forall i \in I, k \in \mathcal{T} \tag{12}$$

The restrictions on decision variables are expressed by the two classical constraints (9, 10, 11, and 12). The developed formulation is a binary mixed program. This program has NK constraints and N^2K^2 decision variables.

3.2   The Flow Formulation Model with Fuzzy Restriction on the Number of Sinks

In most situations from real life, it's assumed that the number of sinks to be located in the network is given in advance by the decision-maker. This assumption expresses, for example, the limited budget to construct the network. Nevertheless, the decision-maker may suggest to construct a network with a sufficient level of coverage and without specifying an exact number of sinks. Certainly, such good level of coverage is achievable only if the the number of the sinks is greater than a specific lower value of p. Otherwise, a redundant coverage, by locating high number of sinks, may ensure the resilience of the network. Accordingly, we may formulate such scenario by locating a number of sinks that belongs to an integer interval $[\![p_1;$

$p_2]\!]$ where p_2 is the maximal number of the sinks to be located. Such situation may be modeled using fuzzy equality as follows:

$$\sum_{i \in N} z_j =_f p \tag{13}$$

The fuzzy equality symbol $=_f$ expresses, as previously explained, the fact that the number of sinks to be placed is taken from the interval $[\![p_1; p_2]\!]$. We can replace the equality Eq. (13) by the two following fuzzy inequalities:

$$\sum_{i \in N} z_j \leq_f p \tag{14}$$

$$\sum_{i \in N} z_j \geq_f p \tag{15}$$

Accordingly, and considering the previous constraints, the problem may be formulated as follows:

$$\text{Min} \sum_{i \in I} \sum_{k \in \mathcal{T}} \sum_{j \in I} \sum_{l \in \mathcal{T}} e_{ijk} \mu_{ijkl} + \sum_{i \in I} \sum_{k \in \mathcal{T}} \sum_{j \in I} e_{ijk} v_{ijk} \tag{16}$$

Subject to

$$\sum_{i \in I} \sum_{k \in \mathcal{T}} q_{ijk} s_{ik} \geq \theta_j, \forall j \in I \tag{17}$$

$$\sum_{i \in I} \sum_{k \in \mathcal{T}} h_k s_{ik} \leq B \tag{18}$$

$$\sum_{j \in I} \sum_{l \in \mathcal{T}} \sum_{k \in \mathcal{T}} u_{jilk} + \sum_{j \in I} \sum_{l \in \mathcal{T}} v_{jil} + \sum_{k \in \mathcal{T}} = \sum_{j \in I} \sum_{l \in \mathcal{T}} \sum_{k \in \mathcal{T}} u_{ijlk} + \sum_{j \in I} \sum_{k \in \mathcal{T}} v_{ijk} + w_i, \forall i \in I \tag{19}$$

$$\sum_{j \in I} \sum_{l \in \mathcal{T}} \mu_{ijkl} + \sum_{j \in N} v_{ijl} \leq NK s_{ik}, \forall i \in I, k \in \mathcal{T} \tag{20}$$

$$\sum_{i \in I} \sum_{k \in \mathcal{T}} v_{ijk} \leq NK z_j, \forall j \in I \tag{21}$$

$$w_i \leq NK y_i, \forall i \in I \tag{22}$$

$$\sum_{i \in N} z_j \leq_f p \tag{23}$$

$$\sum_{i \in N} z_j \geq_f p \tag{24}$$

$$\mu_{ijkl}, \forall i,j \in I, k,l \in \mathcal{T} \tag{25}$$

$$\nu_{ijk}, \forall i,j \in I, k \in \mathcal{T} \tag{26}$$

$$y_i, s_{ik} \in \{0, 1\} \forall i \in I, k \in \mathcal{T} \tag{27}$$

3.3 The Flow Formulation Model with Fuzzy Restriction on Budget

The expected budget to cover the total cost of sensors to be deployed may be fixed by the decision-maker. However, in some cases, the decider may allow an approximate budget. This can be formulated as follows:

$$\sum_{i \in \mathcal{I}} \sum_{k \in \mathcal{T}} h_k s_{ik} \underset{f}{\leq} B \tag{28}$$

This fuzzy inequality means that the total expected budget may belong to an interval $[H_1, H_2]$. Thereby, the single-commodity network with fuzzy budget is given as follows:

$$\text{Min} \sum_{i \in I} \sum_{k \in \mathcal{T}} \sum_{j \in I} \sum_{l \in \mathcal{T}} e_{ijk} \mu_{ijkl} + \sum_{i \in I} \sum_{k \in \mathcal{T}} \sum_{j \in I} e_{ijk} \nu_{ijk} \tag{29}$$

Subject to

$$\sum_{i \in I} \sum_{k \in \mathcal{T}} q_{ijk} s_{ik} \geq \theta_j, \forall j \in I \tag{30}$$

$$\sum_{i \in \mathcal{I}} \sum_{k \in \mathcal{T}} h_k s_{ik} \underset{f}{\leq} H \tag{31}$$

$$\sum_{j \in I} \sum_{l \in \mathcal{T}} \sum_{k \in \mathcal{T}} u_{jilk} + \sum_{j \in I} \sum_{l \in \mathcal{T}} v_{jil} + \sum_{k \in \mathcal{T}} = \sum_{j \in I} \sum_{l \in \mathcal{T}} \sum_{k \in \mathcal{T}} u_{ijlk} + \sum_{j \in I} \sum_{k \in \mathcal{T}} \nu_{ijk} + w_i, \forall i \in I \tag{32}$$

$$\sum_{j \in I} \sum_{l \in \mathcal{T}} \mu_{ijkl} + \sum_{j \in N} v_{ijl} \leq NK s_{ik}, \forall i \in I, k \in \mathcal{T} \tag{33}$$

$$\sum_{i \in I} \sum_{k \in \mathcal{T}} \nu_{ijk} \leq NK z_j, \forall j \in I \tag{34}$$

$$w_i \leq NK y_i, \forall i \in I \tag{35}$$

$$\sum_{j \in N} z_j = p \tag{36}$$

$$\mu_{ijkl}, \forall i,j \in I, k,l \in \mathcal{T} \tag{37}$$

$$\nu_{ijk}, \forall i,j \in I, k \in \mathcal{T} \tag{38}$$

$$y_i, s_{ik} \in \{0, 1\} \forall i \in I, k \in \mathcal{T} \tag{39}$$

3.3.1 Overview on Fuzzy Mathematical Programming

The first mathematical models to formulate real-life problems are deterministic. Researchers consider deterministic and certain situations. However, several situations from engineering field are uncertain and input data are not known in advance. As the probability theory is no longer sufficient to model such situations, a new modeling paradigm based on fuzzy logic was emerged thanks to Lotfi Zadeh since 1965 (Zadeh, 1965). This new concept of logic was intended to help dealing with fuzzy environment.

Fuzzy set theory was then built on fuzzy logic considering fuzzy sets. Fuzzy sets are based on membership function that takes values from the interval [0, 1]. Rapidly, several researchers applied fuzzy set theory in operational research to model fuzzy relationships between various decision components (Zimmermann, 1985).

The authors of (Kaufmann, 1988) introduced "the experton" to describe subjectivity knowledge and uncertain data from one or several experts in computer science (Guiffrida & Nagi, 1998). The objective of the fuzzy theory is to formulate models able to capture uncertainty in input factors. It was agreed that when input factors (data) are expressed as linguistic terms, the fuzzy modeling remains the most suitable formulation to be considered. In fact, following this, fuzzy modeling is a strong tool to cope with uncertainty (Zadeh, 1978). Many applications from business, engineering, health science, etc. are now modeled using fuzzy paradigms. Especially, mathematical programming is a good example of fields in which fuzzy set theory is heavily applied (Zimmermann, 1985). Besides, an interesting review for application for modeling WSN can be found in (Akyildiz et al., 2002).

3.4 A Parametric Solution Approach

The linear mixed-integer programs previously presented are fuzzy mixed-integer linear programs. To the best of our knowledge no method can solve the problem in its original form. Transforming this type of models into a set of crisp problem is a way to deal with them. When converting fuzzy problems to crisp problems, the aim is to solve the crisp problem using exact or heuristic algorithms. The obtained results yield a series of fuzzy program solutions. Reference (Verdegay, 1982) gives a good description of these approaches. The parametric approach, known as α−cut method, transforms the fuzzy models into crisp formulations. The mentioned approach contains two phases. First, the initial fuzzy mixed-integer program to be solved is transformed to a set of non-fuzzy (or crisp) models using α−cuts where α belongs to the interval $[0, 1]$. The decision-maker's tolerance for the given fuzzy constraint is represented by this parameter. Afterward, each crisp program is solved using either exact or heuristic methods (simplex, B&B, heuristics, etc.). The results obtained for different values of α represent a set of solutions to the fuzzy model.

3.4.1 Solving the Flow Formulation with Restriction on the Number of Sinks

Th model for flow formulation with restriction on the number of sinks proposed previously has the two fuzzy restrictions expressed by the two constraints (23) and (24). Let's take the assumption that the decision-maker allows some flexibility in these constraints that is represented, respectively, by the values $p + \delta_r$ and $p + \delta_l$. Let $S = \sum_{j \in I} z_j$, and then, the inequalities (23) and (24) can be expressed, respectively, by the following membership functions:

$$\lambda_r(S) = \begin{cases} 1 \text{ if } S \leq p \\ 1 - \dfrac{S-p}{\delta_r} \text{ if } p \leq S \leq p + \delta_r \\ 0 \text{ if } S > p + \delta_r \end{cases} \tag{40}$$

And

$$\lambda_l(S) = \begin{cases} 1 \text{ if } S \geq p \\ 1 - \dfrac{p-S}{\delta_l} \text{ if } p - \delta_l \leq S \leq p \\ 0 \text{ if } S < p - \delta_l \end{cases} \tag{41}$$

The associated α−cuts are:

$$\lambda_r(S) \geq \alpha, \forall \alpha \in [0, 1]$$

$$\lambda_l(S) \geq \alpha, \forall \alpha \in [0, 1]$$

Accordingly, the α−SLRP (Sink Location and Routing Problem) formulation is given by:

Min Eq. (16)

Subject to:

Eqs. (17, 18, 19, 20, 21, and 22)

$$\sum_{i \in N} z_j \le p + \delta_r(1 - \alpha), \forall \alpha \in [0, 1] \qquad (42)$$

$$\sum_{i \in N} z_j \ge p - \delta_l(1 - \alpha), \forall \alpha \in [0, 1] \qquad (43)$$

The model described above is a linear program that can be solved with standard solvers (such as CPLEX, Gurobi, etc.). By giving values from 0 to 1 with a step of 0.1, to the parameter α, a set of solutions is generated. The latter is also a basis of solution for the transformed fuzzy program.

3.4.2 Solving the Flow Formulation Model with Fuzzy Restriction on Budget

In the model version with fuzzy budget, the fuzziness comes from the constraint (31). For this, we take the assumption that the decision-maker allows the violation of this constraint with tolerance value $H + \delta_h$. Let $B = \sum_{i \in I} \sum_{k \in \mathcal{T}} h_k s_{ik}$,and then, the fuzzy inequality (31) may be expressed by the following membership function:

$$\sigma_h(B) = \begin{cases} 1 \text{ if } B \le H \\ 1 - \dfrac{B - H}{\delta_h} \text{ if } H \le B \le H + \delta_h \\ 0 \text{ if } B > H + \delta_h \end{cases} \qquad (44)$$

And the associated α−cuts is:

$$\sigma_r(B) \ge \alpha, \forall \alpha \in [0, 1]$$

Thereby, the α−SLRP with fuzzy budget is modeled as follows:
Min Eq. (29)
Subject to:
Eqs. (30, 32, 33, 34, 35, 36, 37, 38, and 39)

$$\sum_{i \in I} \sum_{k \in \mathcal{T}} h_k s_{ik} \le H + \sigma_h(1 - \alpha), \forall \alpha \in [0, 1] \qquad (45)$$

As stated in the previous subsection, the generated solutions by solving the above set of crisp models constitute a solution for the model with fuzzy budget constraint.

3.4.3   An Alternative p-Median Formulation

We present in this subsection the Sink Location and Routing Problem (SLRP) as a p-median (El Hamdi et al., 2019). This model is based on formulation proposed by (Güney et al., 2012): We consider the following parameters:

e_{ijk}: The necessary unit energy for transmitting a unit data from the sensor type k located at node i to the sink located at node j

   We consider the following decision variables:

δ_{ijk}: a 0–1 decision variable, equal 1 if a sensor type k at node i is assigned to a sink at node j 0 otherwise

z_j: a 0–1 decision variable, equal 1 in case a sink is located at node j and 0 otherwise

$$\text{Min} \sum_{i \in S} \sum_{k \in K} \sum_{j \in N} e_{ijk} \delta_{ijk} \tag{46}$$

Subject to:

$$\sum_{j \in N} \delta_{ijk} = 1, \forall i \in S, k \in K_i \tag{47}$$

$$\delta_{ijk} \leq z_j, \forall i \in S, k \in K_i, j \in N \tag{48}$$

$$\sum_{j \in N} z_j = p \tag{49}$$

$$\delta_{ijk}, z_j \in \{0, 1\}, \forall i \in S, k \in K_i, j \in N \tag{50}$$

The objective function (46) aims to minimize the total energy for transmitting data flows from different types of sensors to sinks. The constraint (47) guarantees to assign each sensor to exactly one sink. The constraint (48) ensures the assignment of sensors to located sinks. The constraint (49) ensures that the given number of sinks is opened.

3.5   *Locating Sinks Under Fuzzy Constraints*

As stated previously, in most situations from real life, it's assumed that the number of sinks to be located is given in advance by the decision-maker. This restriction reflects the budget limitation to construct the network. Nevertheless, the decision-maker may suggest to construct a network with a sufficient level of coverage and without specifying an exact number of sinks. $[p_{min}; p_{max}]$ where p_{max} is the maximal number of the sinks to be located and p_{min} is the lower bound. Thus, the we formulate the Sink Location and Routing Problem with fuzzy number of sinks as follows:

Min Eq. (46)
Subject to:
Eqs. (47, 48, and 50)

$$\sum_{j \in N} z_j = p \atop f \tag{51}$$

This Eq. (51) is equivalent to the two following inequalities:

$$\sum_{j \in N} z_j \geq p \atop f \tag{52}$$

$$\sum_{j \in N} z_j \leq p \atop f \tag{53}$$

4 Observability and Sensor Location Problem on Highway Segments

To solve the problem of road congestion, researchers must observe the traffic conditions on the highways. To collect traffic data on highways, the sensors are used. Nerveless, to minimize cost installations and maximize the information collected, sensors must be placed efficiently. In this part we present an approach to study the problem of observability on motorway segments using the dynamics of linearized traffic on steady-state flows (Contreras et al., 2015).

4.1 Problem Statement

For our mathematical modeling, we will use macroscopic traffic flow models describing the evolution of the position of vehicle in terms of the average speed and the density of cars. So, we consider an interval $I = [a, b]$ of $\mathbf{R}(a < b)$ modeling a road. The average velocity $v(t, x)$ and density $\rho(t, x)$ depend on the time t and on $x \in I$ (Garavello & Piccoli, 2006). For this end, we consider the scalar model proposed by Richards in 1956 and Lighthill and Whitham in 1955. By using the conservation of cars the model is given by the Lighthill-Whitham-Richards (LWR) equation:

$$\rho_t + f(\rho)_x = 0$$

Note that the traffic flux (units of cars/time) $q = f(\rho, v)$ is given by ρv. The average speed v is in general a function that depends on the density. And we have the following assumptions:

(A1) $f \in C^2(\mathbb{R})$.

(A2) f is strictly concave.

(A3) $f(0) = f(\rho_{max}) = 0$.

We call fundamental diagram the flux as function of the density. Various fundamental diagrams describe the velocity function $v = v(\rho)$; thus, we obtain the flux by multiplying the density ρ. In the case when v is a linear function of the density, a simple fundamental diagram is given by

$$v(\rho) = v_{max} \left(1 - \frac{\rho}{\rho_{max}} \right)$$

Many another fundamental diagram was considered in the literature (Garavello & Piccoli, 2006). Supported by experimental data from the Lincoln tunnel in New York, Greenberg proposed the velocity function

$$v(\rho) = v_0 \log \left(\frac{\rho_{max}}{\rho} \right)$$

where v_0 is a positive constant. In this case $v(\rho_{max}) = 0$, while v is unbounded when $\rho \to 0^+$.

4.2 Nonlocal Conservation Law Model

Nonlocal conservation laws have been used in recent years, and for modeling traffic flow and other manufacturing process, more details can be found in (Armbruster et al., 2006). In order to study the observability of the system, we will consider the nonlocal conservation law model given by the nonlinear transport equation:

$$\rho_t(t, x) + (\rho(t, x)\lambda(W(t)))_x = 0, \quad t \in (0, +\infty), \quad x \in (0, 1) \tag{54}$$

where the WIP (work in process) is given by

$$W(t) = \int_0^1 \rho(t, x)dx$$

The nonlocalness of the conservation law yields in the velocity $\lambda(W(t))$ which depends on the integral of $\rho(t, x)$ in space. Typical velocity function is suggested from queueing theory and given by

$$\lambda(s) = \frac{1}{1 + s}, \quad s \in [0, +\infty)$$

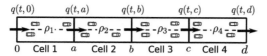

$q(t,0)$ $q(t,a)$ $q(t,b)$ $q(t,c)$ $q(t,d)$

0   Cell 1 a Cell 2 b Cell 3 c Cell 4 d

Fig. 1 Space discretization [38]

In semiconductor manufacturing modeling, the previous velocity function is widely used. Its main feature is decreasing in the case of increasing of the total WIP of the system.

4.2.1 The Microscopic Model for Nonlocal Conservation Law

In this part we extend the discretization method used in (Contreras et al., 2015) to the nonlocal model, and the idea is to consider ODE approximation of nonlocal LWR model for traffic road. We denote $\rho_i(t)$ the average density of the highway cell i (see Fig. 1). Thus, the obtained system is the average densities describing the change in time of the, ρ_i 's. Space discretization of Eq. (54) for the traffic flow on a network link is presented in Fig. 1. The length of the network link is assumed to be uniform and equal to ℓ, and then the ordinary differential equation for the traffic flow on a network link is given by the following equation:

$$\frac{d\rho(t)}{dt} = \frac{f_{in}(t) - f_{out}(t)}{\ell}$$

$$q(t, i) = f_{out}(t) = \rho_i(t) \left(\frac{1}{1 + \ell \rho_i(t)} \right)$$

Thus, for each arc on the traffic network, we obtain the ODE model k:

$$\frac{d\rho_i(t)}{dt} = \frac{1}{\ell_i} \left(f_{in_i}(t) - \rho_i(t) \left(\frac{1}{1 + \ell_i \rho_i(t)} \right) \right)$$

Space discretization

$$\dot{\rho}_1 = h_1(\rho_1, \rho_2, \ldots, \rho_n) = \frac{(f_{in} - q(t, 1))}{l}$$

$$\dot{\rho}_2 = h_2(\rho_1, \rho_2, \ldots, \rho_n) = \frac{(q(t, 1) - q(t, 2))}{l}$$

$$\vdots$$

$$\dot{\rho}_n = h_n(\rho_1, \rho_2, \ldots, \rho_n) = \frac{(q(t, n-1) - q(t, n))}{l}$$

This system is clearly nonlinear. In order to linearize our system, we begin by finding the equilibrium point. To find the equilibrium point, we solve the equation

$$f_{in_i}(t) - \rho_i(t)\left(\frac{1}{1 + \ell_i \rho_i(t)}\right) = 0$$

which gives

$$\rho_i^* = \left(\frac{f_{in_i}}{1 + f_{in_i}\ell_i}\right)$$

$$\rho_1^* = \left(\frac{f_{in_0}}{1 + f_{in_0}\ell_i}\right)$$

Therefore, for i=2,3,...

$$\rho_i^* = \left(\frac{\rho_{i-1}^*}{1 + \rho_{i-1}^*(\ell_{i-1} - \ell_i)}\right)$$

that densities ρ_1, ρ_2, ρ_3, ρ_4 do not change with respect to time.

$$0 = \frac{1}{l_1} f_{in} - \frac{1}{l_1} v_f \rho_{1_{eq}}\left(1 - \frac{\rho_{1_{eq}}}{\rho_{max}}\right)$$

$$0 = \frac{1}{l_2} v_f \rho_{1_{eq}}\left(1 - \frac{\rho_{1_{eq}}}{\rho_{max}}\right) - \frac{1}{l_2} v_f \rho_{2_{eq}}\left(1 - \frac{\rho_{2_{eq}}}{\rho_{max}}\right)$$

$$0 = \frac{1}{l_3} v_f \rho_{2_{eq}}\left(1 - \frac{\rho_{2_{eq}}}{\rho_{max}}\right) - \frac{1}{l_3} v_f \rho_{3_{eq}}\left(1 - \frac{\rho_{3_{eq}}}{\rho_{max}}\right)$$

$$0 = \frac{1}{l_4} v_f \rho_{3_{eq}}\left(1 - \frac{\rho_{3_{eq}}}{\rho_{max}}\right) - \frac{1}{l_4} v_f \rho_{4_{eq}}\left(1 - \frac{\rho_{4_{eq}}}{\rho_{max}}\right)$$

4.3   Observability of the Traffic Flow System

In this part we recall the notion of observability and controllability. Indeed, the observability expresses the capacity of a sensor to collect the information for estimating different states of the system. Conversely, the controllability expresses the capacity of an actuator to control the states of the system. Consider the following linear system:

$$\dot{z} = Az + Bu \tag{55}$$

$$y = Cz \tag{56}$$

where A, B, and C are constant matrices of appropriate dimension.

The pair (A, C) is said to be observable if any initial state $z(0)$ is uniquely determined from the output $y(k)$ and input sequence $u(k)$, for $k = 0, 1, 2, \cdots, n$.

In order to study the observability, we linearize a nonlinear system about an equilibrium point by using the Jacobian of f. Consequently, we get a linear state-space system for control deviations and state: $\Delta x = A\Delta x$, where since $\dot{\rho}_{eq_i} = 0$, a new variable, Δx, is defined.

$$\Delta x = \rho_i - \rho_{eq_i}, \quad i = 1,2,\ldots,n.$$
$$\dot{\Delta x} = \dot{\rho}_i, \quad i = 1,2,\ldots,n.$$

These new variables give the following equation:

$$\begin{bmatrix} \dot{\Delta x}_1 \\ \dot{\Delta x}_2 \\ \vdots \\ \Delta x_n \end{bmatrix} = \begin{bmatrix} \dfrac{-1}{l(1+l\rho_1)} & 0 & \cdots & 0 & 0 \\ \dfrac{1}{l(1+l\rho_1)} & \dfrac{-1}{l(1+l\rho_2)} & \cdots & 0 & 0 \\ \vdots & & \ddots & \vdots & \vdots \\ 0 & 0 & \cdots & \dfrac{1}{l(1+l\rho_n)} & \dfrac{-1}{l(1+l\rho_n)} \end{bmatrix} \times \begin{bmatrix} \Delta x_1 \\ \Delta x_2 \\ \vdots \\ \Delta x_n \end{bmatrix}$$

Thus to obtain the placement of actuator, we use the following theorem.

Theorem 1

The pair (A, C) is observable if the observability matrix given by (Preumont, 1997)

$$U_O = \begin{bmatrix} C \\ CA \\ CA^2 \\ \vdots \\ CA^{n-1} \end{bmatrix}$$

has rank n, i.e., full column rank. Equation (55) describes the linearized and discretized dynamics for average densities in n different cells about an equilibrium point. Measurements of the states are represented in the form shown below.

$$y = Cz$$

where $y \in \mathbb{R}^p$, $C \in \mathbb{R}^{p \times n}$, and $z \in \mathbb{R}^n$. y is a column of measurements and p is the number of measurements.

The matrix C is used to identify how sensors are placed. As an example, when there are four cells, $n = 4$, and two sensors, $p = 2$, measure z_1 and z_3, the matrix C is

$$C = \begin{bmatrix} 1 & 0 & 0 & 0 \\ 0 & 0 & 1 & 0 \end{bmatrix}$$

Cases of interest are when $p < n$. This is when not all of the states of the system are measured. Otherwise, when $p = n$, measurements of all cells are taken.

5 Conclusion

The use of ICT technologies in transportation has enabled to design smart mobility systems. In fact, the growing use of technological innovations, especially Internet of Things (Iot), has offered many opportunities to develop intelligent and sustainable solutions on the transportation sector. In the context of Industry 4.0, smart transportation or transportation 4.0 paradigm is intended to be a competitive, flexible, safer, and sustainable alternative for the current mobility systems. Such intelligent mobility systems are based on data from the network collected using a network of wireless sensors. For this reason, the wireless sensor placement problem is an important problem that has attracted the attention of both practitioners and academics. In this paper, we studied this problem using fuzzy linear programming approach and from observability of the traffic flow standpoint. A first model is the single-commodity network flow with fuzzy constraint on the number of the sinks to be located. The second one is the Sink Location and Routing Problem with fuzzy restrictions on the number of sinks to be opened.

All these fuzzy formulations were converted to crisp models using the well-known $\alpha-$cut method. The generated set of crisp models are solved to optimality using CPLEX solver.

Furthermore, we studied the observability and sensor location problem on highway segments. We used the macroscopic traffic flow model to formulate the evolution of the position of a given vehicle in the network. The proposed model is then extended to the nonlocal conservation case using a microscopic model.

References

Akyildiz, I. F., Weilian, S., Sankarasubramaniam, Y., & Cayirci, E. (2002). A survey on sensor networks. *IEEE Communications Magazine, 40*(8), 102–114.

Armbruster, D., Marthaler, D. E., Ringhofer, C., Kempf, K., & Jo, T.-C. (2006). A continuum model for a re-entrant factory. *Operations Research, 54*(5), 933–950.

Bag, S., Yadav, G., Wood, L. C., Dhamija, P., & Joshi, S. (2020). Industry 4.0 and the circular economy: Resource melioration in logistics. *Resources Policy, 68*, 101776.

Benevolo, C., Dameri, R.P., & D'auria, B. (2016). Smart mobility in smart city. In *Empowering organizations* (pp. 13–28). Springer.

Bottero, M., Chiara, B. D., & Deflorio, F. P. (2013). Wireless sensor net- works for traffic monitoring in a logistic centre. *Transportation Research Part C: Emerging Technologies, 26*, 99–124.

Bubelíny, O., Kubina, M., & Varmus, M. (2021). Railway stations as part of mobility in the smart city concept. *Transportation Research Procedia, 53*, 274–281.

Buyya, R., & Dastjerdi, A. V. (2016). *Internet of Things: Principles and paradigms.* Elsevier.

Contreras, S., Kachroo, P., & Agarwal, S. (2015). Observability and sensor placement problem on highway segments: A traffic dynamics-based approach. *IEEE Transactions on Intelligent Transportation Systems, 17*(3), 848–858.

Daws, R.. (2021, June). Volvos trucks and cars will communicate to improve safety. [Online] *iottechnews*. Available at: https://iottechnews.com/news/2018/may/08/volvo-trucks-cars-improve-safety/. Accessed 21 June 2021.

Dima, S. M., Panagiotou, C., Tsitsipis, D., Antonopoulos, C., Gi-alelis, J., & Koubias, S. (2014). Performance evaluation of a wsn system for distributed event detection using fuzzy logic. *Ad Hoc Networks, 23,* 87–108.

El Hamdi, S., Oudani, M., Abouabdellah, A., & Sebbar, A. (2019). Fuzzy approach for locating sensors in industrial internet of things. *Procedia Computer Science, 160,* 772–777.

El Raoui, H., Oudani, M., & Alaoui, A. E. H. (2020). Coupling soft computing, simulation and optimization in supply chain applications: Review and taxonomy. *IEEE Access, 8,* 31710–31732.

Elhoseny, M., Tharwat, A., Yuan, X., & Hassanien, A. E. (2018). Optimizing k-coverage of mobile wsns. *Expert Systems with Applications, 92,* 142–153.

Faye, S., & Chaudet, C. (2015). Characterizing the topology of an urban wireless sensor network for road traffic management. *IEEE Transactions on Vehicular Technology, 65*(7), 5720–5725.

Garavello, M., & Piccoli, B. (2006). *Traffic flow on networks* (Vol. 1). American Institute of Mathematical Sciences Springfield.

Guiffrida, A. L., & Nagi, R. (1998). Fuzzy set theory applications in production management research: A literature survey. *Journal of Intelligent Manufacturing, 9*(1), 39–56.

Güney, E., Necati Aras, İ., Altınel, K., & Ersoy, C. (2012). Efficient solution techniques for the integrated coverage, sink location and routing problem in wireless sensor networks. *Computers & Operations Research, 39*(7), 1530–1539.

Hashish, S., Tawalbeh, H., et al. (2017). Quality of service requirements and challenges in generic wsn infrastructures. *Procedia Computer Science, 109,* 1116–1121.

Jiang, J., Wang, H., Xiangwei, M., & Guan, S. (2020). Logistics industry monitoring system based on wireless sensor network platform. *Computer Communications, 155,* 58–65.

Karami, Z., & Kashef, R. (2020). Smart transportation planning: Data, models, and algorithms. *Transportation Engineering, 2,* 100013.

Kaufmann, A. (1988). Theory of expertons and fuzzy logic. *Fuzzy Sets and Systems, 28*(3), 295–304.

Mohamed, S. M., Hamza, H. S., & Saroit, I. A. (2017). Coverage in mobile wireless sensor networks (m-wsn): A survey. *Computer Communications, 110,* 133–150.

Ning, L., Cheng, N., Zhang, N., Shen, X., & Mark, J. W. (2014). Connected vehicles: Solutions and challenges. *IEEE Internet of Things Journal, 1*(4), 289–299.

Oudani, M. (2021). A simulated annealing algorithm for intermodal transportation on incom- plete networks. *Applied Sciences, 11*(10), 4467.

Oudani, M., & Zkik, K. (2019). Fuzzy single-commodity model in wireless sensor net- works. *Procedia Computer Science, 160,* 797–802.

Patil, H. K., & Chen, T. M.. (2013). Wireless sensor network security. In *Computer and information security handbook* (pp. 301–322). Elsevier.

Peixoto, J. P. J., & Costa, D. G. (2017). Wireless visual sensor networks for smart city applications: A relevance-based approach for multiple sinks mobility. *Future Generation Computer Systems, 76,* 51–62.

Porru, S., Misso, F. E., Pani, F. E., & Repetto, C. (2020). Smart mobility and public transport: Opportunities and challenges in rural and urban areas. *Journal of Traffic and Transportation Engineering (English Edition), 7*(1), 88–97.

Preumont, A. (1997). *Vibration control of active structures* (Vol. 2). Springer.

Qiao, B. (2018). *Revenue Management pour les prestataires de services logistiques dans l'internet physique: les transporteurs de fret comme cas.* PhD thesis, Paris Sciences et Lettres.

Rafael Asorey-Cacheda, A.-J., Garcia-Sanchez, F. G.-S., & García-Haro, J. (2017). A survey on non-linear optimization problems in wireless sensor networks. *Journal of Network and Computer Applications, 82,* 1–20.

Sivakumar, P., & Radhika, M. (2018). Performance analysis of leach-ga over leach and leach-c in wsn. *Procedia Computer Science, 125*, 248–256.

Song, Z., Lazarescu, M. T., Tomasi, R., Lavagno, L., & Spirito, M. A. (2014). High-level internet of things applications development using wireless sensor networks. In *Internet of Things* (pp 75–109). Springer.

Souissi, I., Azzouna, N. B., & Said, L. B. (2019). A multi-level study of information trust models in wsn-assisted iot. *Computer Networks, 151*, 12–30.

Tao, D., Shouning, Q., Liu, F., & Wang, Q. (2015). An energy efficiency semi-static routing algorithm for wsns based on hac clustering method. *Information Fusion, 21*, 18–29.

Thangaramya, K., Kanagasabai Kulothungan, R., Logambigai, M. S., Ganapathy, S., & Kannan, A. (2019). Energy aware cluster and neuro-fuzzy based routing algorithm for wireless sensor networks in iot. *Computer Networks, 151*, 211–223.

Touati, Y., Daachi, B., & Arab, A. C. (2017). *Energy management in wireless sensor networks.* Elsevier.

Verdegay, J. L. (1982). Fuzzy mathematical programming. *Fuzzy Information and Decision Processes, 231*, 237.

Zadeh, L. A. (1965). Information and control. *Fuzzy Sets, 8*(3), 338–353.

Zadeh, L. A. (1978). Fuzzy sets as a basis for a theory of possibility. *Fuzzy Sets and Systems, 1*(1), 3–28.

Zimmermann, H.-J. (1985). Applications of fuzzy set theory to mathematical programming. *Information Sciences, 36*(1–2), 29–58.

Barriers in Smart Green Resilient Lean Manufacturing: An ISM Approach

Imane Benkhati, Fatima Ezahra Touriki, and Said El Fezazi

1 Background Literature

1.1 Smart Manufacturing

The agency of the US Department of Commerce defines smart manufacturing (SM) as 'a fully integrated and collaborative manufacturing system that responds in real time to meet the changing demands and conditions in the factory, supply network and customer need'. Authors across the literature agree that SM can revolutionize existing manufacturing practices, leading to the enhancement of productivity, quality, delivery and cost (Kang et al., 2016). This new mode of manufacturing can be achieved using several technologies like cyber-physical systems (CPS), the Internet of Things (IoT), big data analytics (BDA), additive manufacturing (AM), artificial intelligence (AI), augmented and virtual reality (AR, VR) and cloud computing (Kusiak, 2018).

AI enables manufacturing systems to rely less on human involvement to make different decisions in real time. BDA plays an important role in maintaining the equipment maintenance, precise fault detection, fault precision and cost elimination (Tao et al., 2018). Data and information regarding temperature, pressure and vibrations can be easily obtained by the means of IoT (Zheng et al., 2018). AM helps increase flexibility and reduce time and cost for machine tool replacements as well as for raw materials (Matos & Jacinto, 2019). VR and AR were found to reduce the cost

I. Benkhati (✉) · F. E. Touriki
LMPEQ, ENSA-Safi, Cadi Ayyad University, Marrakech, Morocco
e-mail: Imane.benkhati@ced.uca.ma; f.touriki@uca.ma

S. El Fezazi
LaPSSII, EST-Safi, Cadi Ayyad University, Marrakech, Morocco

of design and prototyping, besides their employment for training purposes (Webel et al., 2013). CPS provides both monitoring and control from anywhere and anytime (Yu et al., 2015). In a nutshell SM offers great opportunities in addressing manufacturing current challenges and providing smart solutions (Kusiak, 2018).

1.2 Green Manufacturing

Melnyk and Smith (1996) define green manufacturing (GM) as 'a system that integrates product and process design issues with issues of manufacturing planning and control in such a matter to identify, quantify assess and manage the flow of environmental waste with the goal of reducing and ultimately minimizing environmental impact while trying to maximize resource efficiency'. GM is the outcome of stringent environmental regulations, resources scarcity, global warming, waste management issues and higher global awareness (Paul et al., 2014).

The adoption of green manufacturing will benefit manufacturers from a multitude of advantages. First, it will not only ensure compliance with regulations, but it will also result in efficiency by utilizing less water, energy and raw materials. Furthermore, it will reduce safety expenses and improve corporate image. Finally, GM is an opportunity to gain a competitive advantage and shift towards a cleaner and more sustainable manufacturing (Lee et al., 2019; Diekmann & Preisendörfer, 2003; Govindan et al., 2015).

1.3 Resilient Manufacturing

Kusiak (2020) defines resilient manufacturing (RM) as 'a mode of manufacturing that assesses vulnerability to the unexpected disruptions and mitigates their impact accordingly'. It is characterized by the ability to recover from an 'undesired state and to a desired one'. Heinicke (2014) highlights the importance of resilience in manufacturing, reporting that disruptive events like machinery troubleshooting and breakdowns can result in the entire or the partial loss of production.

There are several factors that impact manufacturing resilience. We mainly distinguish between internal disruptions such as manufacturing troubleshooting, safety incidents and operator strikes and external ones like shortage of materials, energy and natural disasters (Amjad et al., 2020).

1.4 Lean Manufacturing

Lean manufacturing (LM) is a mode of manufacturing that focuses on waste elimination and value creation. It defines customer value, defines value stream, makes it flow, establishes pull and strives for excellence (Womack & Jones,

1996). LM is the outcome of a fluctuating and competitive business environment, with the aim to maximize resource utilization through waste minimization (Sundar et al., 2014). LM is multi-dimensional; hence it utilizes various practices, like 'quality systems, just in time, cellular manufacturing, supplier management, employee involvement, quality circles, levelled scheduling, Kanban, visual control, SMED...' etc. (Shah & Ward, 2003).

Through the literature, various benefits are reported through the application of LM. First, product quality is improved, lead and cycle times are reduced and costs are decreased. Second, labour and resources are re-allocated in a more efficient way. Finally, inventories are reduced and flexibility is increased (Pavnaskar et al., 2003). All in all, LM when implemented successfully can result in a more robust process and improved knowledge management (Melton, 2005).

2 Research Gaps

While smart, green, resilient and lean paradigms are crucial components for a successful and profitable manufacturing system, passably little effort has been made to examine the potential linkages between integrating the four concepts (Amjad et al., 2021). Although SGRLM represents a smart strategy that will take advantage of the benefits of each paradigms, literature suggests that its implementation comes with multiple barriers (Rajesh, 2018; Ruben et al., 2016; Klein et al., 2018; Singh et al., 2020). A literature gap exists in the identification of barriers relevant to SGRLM; the dynamics of these barriers need to be examined for the ease of their removal. For all we know no previous study has identified the set of barriers which need to be removed for SGRLM. As a result, these literature gaps have formulated the direction of this study.

3 Barriers to SGRLM

Through the review of the literature, we were able to identify 12 barriers that hinder SGRLM implementation. The description of the 12 barriers is elaborated below.

3.1 Lack of Expertise and Training Programs (B1)

Lack of qualified profiles may hinder the implementation of SGRLM, since this new mode of manufacturing will require totally new competencies and skills. Digitalization will increase complexity and abstraction; therefore employees are called to develop advanced problem-solving skills (Peillon & Dubruc, 2019). On the other hand, GM and LM would require staff to have proper knowledge and special training programs (Sindhwani et al., 2019; Mollenkopf et al., 2010). Into the bargain, organization personnel are unfamiliar with ways to manage risks and to build resilience (Rajesh, 2018). Consequently, there is work to be done in the arena of competence buildings when shifting to SGRLM.

3.2 Unsupportive Organization Culture and Resistance to Change (B2)

'For the success of any new initiative, a change in organizational culture needs to take place' (Rimanoczy & Pearson, 2010). Therefore, unsupportive organization culture can hinder the success of SGRLM. Digitalization represents a radical change in key business operations, processes and management concepts along with organizational structures. Consequently, the organization culture should also follow this change (Peillon & Dubruc, 2019). Companies seem to be still unaware of the benefits that come along with green and lean manufacturing, according to Shibin et al. (2016). Further, manufacturers are currently lacking risk management culture (Ruben et al., 2016). Moreover, with the shift towards SGRLM, employees are expected to show resistance, since personnel is often unwilling to accept changes that occur in daily routine (Sindhwani et al., 2019).

3.3 Lack of Management Involvement (B3)

The lack of support of top management has been found to be a leading barrier to the successful implementation of SGRLM practices. Without the commitment of senior management, the implementation of any process will not see the light of the day, since it lacks active support from top management (Mudgal et al., 2010) Green and lean practices require a high level of management participation (Mudgal et al., 2010; Ruben et al., 2016). Furthermore, managers need to be committed to not only providing necessary resources to invest in smart technologies but also continuously motivating employees to adopt new practices (Zhang et al., 2019).

3.4 Lack of Time and Resources (B4)

The shift towards SGRLM is time-consuming, and it requires both technical and human resources. Besides it includes multiple stages of developing, testing and training which can interfere with the daily routine with employees. Therefore inadequate time management and non-optimized resource allocation can hinder the successful implementation of SGRLM (Kumar et al., 2016; Singh et al., 2020; Ghazilla et al., 2015).

3.5 Financial Constraints (B5)

SGRLM will require a high cost of implementation and important investments. The technologies of I4.0 require immense financial resources in both stages of research and development. In addition, the required human resources to support these technologies are highly paid due to skill scarcity (Kamble et al., 2018). On the other hand, the modification of current practices, especially cleaner processes, will require an immense investment. The latter is not always justified and its benefits are not always clear to the top management (Singh et al., 2020). Besides, LM initiatives also require financial support in order to succeed (Mudgal et al., 2010). Therefore, investing in resilience applications is expensive due to the redundancy attribute. A resilient firm may never know what disruptions it has prevented thanks to its investments in resilience; nevertheless, it is ambiguous how to shape the ROI (Pettit et al., 2019).

3.6 Lack of Regulations and Government Support (B6)

The lack of government support and the increase of legal restrictions can hamper the implementation of SGRLM. Environmental regulations and enforcement in particular are still weak; hence, manufacturers don't feel the obligation to invest in the innovative technologies of green practices (Zhang et al., 2019; Mittal & Sangwan, 2014). In addition, digitalization presents itself as a barrier for the law. This digital transformation is still in its infancy, and therefore, in the development phase, there is a 'significant lack of new organizational policies, procedures, global standards, and data sharing protocols' (Kamble et al., 2018; Lohmer & Lasch, 2020).

3.7 Ineffective Communication(B7)

Healthy communication skills have been found to mitigate problems when implementing new policies (Singh et al., 2020). Likewise, for the successful implementation of SGRLM, employees at all levels need to be informed about the various changes that will take place with this shift, and team management and coordination needs to be enhanced. Consequently, the lack of effective communication can seriously hinder the enforcement of SGRLM.

3.8   Lack of Visibility (B8)

The inefficient organizational structure can hinder the flow of information to various levels of the organization (Ghazilla et al., 2015); thus visibility can be hampered and information can be distorted. Furthermore, some suppliers are unwilling to share with factories some valuable and critical information like lead time and capacity capabilities, because they consider data as a competitive advantage (Jones et al., 2009). As a result, both manufacturing vulnerabilities and non-value activities will increase.

3.9   Customer Unawareness and Low Demand (B9)

Many customers are unaware of the benefits escorting 'eco-friendly' products and services (Sindhwani et al., 2019). This can be explained by the fact that customers are price sensitive and aspire for cheaper products, since green comes at an expensive cost (Mittal & Sangwan, 2014; Mittal et al., 2016). This lack of customer involvement will eventually impact the company's green factory. Therefore, pressure from customers to adopt green will push manufacturers to further enhance their green practices (Ruben et al., 2016).

3.10   Lack of Planning (B10)

There is a lack among manufacturers in applying proper planning and setting up clear and well-defined objectives (Singh et al., 2020). Lean and green activities planned in a particular day need to be necessarily done in the exact day and cannot be postponed to another day (Sindhwani et al., 2019). Furthermore, there is a significant lack in contingency plans (Rajesh, 2018). Firms need to improve their contingency plans in order to cope with the unwanted effect of manufacturing uncertainties.

3.11   Lack of Metrics for SGRLM (B11)

Metrics are significantly important to support the successful implementation of any new initiatives (Mudgal et al., 2010). Likewise, for manufacturers to succeed in putting SGRLM in place, they will need to develop appropriate metrics and measurement methods and to take action linking performance measurement to their SGRL practices.

3.12 Technological Constraints (B12)

SGRLM is also faced by technological barriers ranging from security and privacy issues to uncompleted and consistent data. I4.0 platform is escorted with a huge amount of data which makes it vulnerable to cybersecurity threats (Kamble et al., 2018). Into the bargain, implementing newer complex technologies can result in compatibility issues with the existing systems (Mittal & Sangwan, 2014). Further, digitalization must be supported by a reliable technical infrastructure (Peillon & Dubruc, 2019). Finally, data mining and analysis is crucial after the identification of green and lean activities (Sindhwani et al., 2019); hence, data needs to be complete and available in a timely manner in the context of I4.0.

4 Research Methodology

We first identified the set of 12 barriers from the literature, and then we examined the contextual relationships among these barriers using the ISM method. Initially ten experts were contacted via email to elucidate the concept of SGRLM. Only five out of ten expressed an interest in the research. In this group, the industry experts include a production manager, an engineering and project manager, a supply chain manager, a planning and procurement manager and a plant manager. All experts have a long experience of more than 15 years; they are highly skilled in their profession and handy with decision-making. Following the problem analysis and discussion, by the means of a brainstorming session, we opted for the ISM-based modelling approach. Research model is presented in Fig. 1.

5 Interpretive Structural Modelling (ISM)

Interpretive structural modelling (ISM) is 'an interactive learning process that enables the individual or a group of individuals to manage the interrelations between two or more variables at a time, and to summarize the contextual relationships among them without compromising and deviating from the initial properties of the original variables' (Attri et al., 2013). This technique allows for order for any complex problem under consideration (Jharkharia & Shankar, 2005). Various steps are involved in ISM (Kumar et al., 2016):

- Identification of the variables to be studied. In our case the barriers in SGRLM are the variables constituting the system.
- Identification of contextual relationships among variables.
- Establishment of the self-interaction matrix (SSIM).

Fig. 1 Methodology framework used in this study

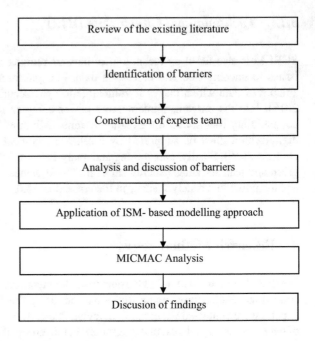

- Establishment of the reachability matrix from the SSIM. It is then checked for transitivity.
- Establishment of level partitions.
- Establishment of directed graph from the final reachability matrix.

5.1 SSIM Development

As mentioned earlier, the teamwork of industrial experts is well conversant with the discussed concepts of SGRLM. For the analysis of the factors, the members established the contextual relationships 'leads to' or 'influences' among the identified set of variables.

We used four symbols to denote the type and direction of relationship between a pair of barriers (i and j):

- 'V – barrier i affects barrier j'.
- 'A – barrier j affects barrier I'.
- 'X – both barriers i and j affect each other'.
- 'O – barriers i and j are unrelated' (Table 1).

Table 1 SSIM of barriers in SGRLM

Barrier	B12	B11	B10	B9	B8	B7	B6	B5	B4	B3	B2
B1	V	V	V	O	O	V	O	A	A	A	X
B2	V	O	O	A	O	V	A	A	A	X	
B3	O	A	O	A	A	V	A	A	A		
B4	O	O	V	O	O	V	O	A			
B5	V	O	O	O	O	O	A				
B6	V	O	O	V	O	O					
B7	X	A	X	O	X						
B8	X	X	A	A							
B9	O	O	O								
B10	O	V									
B11	O										

5.2 Initial Reachability Matrix Formation

Next, we developed the initial reachability matrix from SSIM, where the four symbols '*V, A, X* and *O'* of SSIM were replaced by '1' or '0' in the initial reachability matrix. The rules for this substitution are as follows:

- 'If the (i, j) value in the SSIM is *V*, then the (i, j) value in the reachability matrix becomes *1* and the (j, i) value becomes *0'*.
- 'If the (i, j) value in the SSIM is *A*, then the (i, j) value in the matrix becomes *0* and the (j, i) entry becomes*1'*.
- 'If the (i, j) value in the SSIM is *X*, then the (i, j) value in the matrix becomes *1* and the (j, i) entry also becomes *1'*.
- 'If the (i, j) entry in the SSIM is *O*, then the (i, j) entry in the matrix becomes *0* and the (j, i) entry also becomes *0'*.

5.3 Final Reachability Matrix Formation

By applying the 'transivity rule', we were able to deduce the final reachability matrix from the initial reachability (Tables 2 and 3).

5.4 Level Partitions

The final reachability matrix is used for level partitioning. It is based on 'the comparison of the reachability and the antecedent sets of variables' (Attri et al., 2013). Table 4 indicates clearly that 'Lack of metrics and lack of visibility' are identified at the top level.

Table 2 Initial reachability matrix of barriers in SGRLM

Barrier	B1	B2	B3	B4	B5	B6	B7	B8	B9	B10	B11	B12
B1	1	1	0	0	0	0	1	0	0	1	1	1
B2	1	1	1	0	0	0	1	0	0	0	0	1
B3	1	1	1	0	0	0	1	0	0	0	0	0
B4	1	1	1	1	0	0	1	0	0	1	0	0
B5	1	1	1	1	1	0	0	0	0	0	0	1
B6	0	1	1	0	1	1	0	0	1	0	0	1
B7	0	0	0	0	0	0	1	1	0	1	0	1
B8	0	0	1	0	0	0	1	·1	0	0	1	1
B9	0	1	1	0	0	0	0	0	1	0	0	0
B10	0	0	0	0	0	0	1	0	0	1	1	0
B11	0	0	1	0	0	0	1	1	0	0	1	0
B12	0	0	0	0	0	0	1	1	0	0	0	1

Table 3 Final reachability matrix of barriers in SGRLM

Barrier	B1	B2	B3	B4	B5	B6	B7	B8	B9	B10	B11	B12
B1	1	1	1^a	0	0	0	1	1^a	0	1	1	1
B2	1	1	1	0	0	0	1	0	0	0	0	1
B3	1	1	1	0	0	0	1	0	0	0	0	0
B4	1	1	1	1	0	0	1	1^a	0	1	1^a	1^a
B5	1	1	1	1	1	0	1^a	0	0	0	0	1
B6	0	1	1	0	1	1	1^a	1^a	1	0	0	1
B7	0	0	1^a	0	0	0	1	1	0	1	0	1
B8	1^a	0	1	0	0	0	1	1	0	1^a	1	1
B9	0	1	1	0	0	0	1^a	0	1	0	1^a	1^a
B10	0	0	0	0	0	0	1	0	0	1	1	0
B11	0	1^a	1	0	0	0	1	1	0	1^a	1	1^a
B12	0	0	0	0	0	0	1	1	0	1^a	0	1

[a]Entries indicates the presence of transitivity

5.5 Formation of ISM-Based Model

The final structural model is generated as shown in Fig. 2. We can deduce from Fig. 2 that barrier B6 (lack of regulations and government support) is a highly significant barrier to SGRLM implementation as it forms the base of the ISM hierarchy. This lack of regulations and government support may result in the lack of fund allocation (B5), as governments need to support manufacturers to benefit from incentives and low interest loans to support their SGRLM activities.

As elaborated in the proposed ISM-based model, these financial constraints may further lead to a lack of resources and time allocation in SGRLM (B4).

Table 4 Partitioning of barriers in SGRLM: summary of iterations

Barrier	Reachability set	Antecedent set	Intersection set	Level
B1	*B1, B2, B3, B7, B8, B10, B11, B12*	*B1, B2, B3, B4, B5, B8*	*B1, B2, B3*	*IV*
B2	*B1, B2, B3, B7, B12*	*B1, B2, B3, B4, B5, B6, B9, B11*	*B1, B2, B3*	*II*
B3	*B1, B2, B3, B7*	*B1, B2, B3, B4, B5, B6, B7, B8, B9, B10, B11*	*B1, B2, B3, B7*	*I*
B4	*B1, B2, B3, B4, B7, B8, B10, B11, B12*	*B4, B5*	*B4*	*V*
B5	*B1, B2, B3, B4, B5, B7, B12*	*B5, B6*	*B5*	*VI*
B6	*B2, B3, B5, B6, B7, B8, B9, B12*	*B6*	*B6*	*VII*
B7	*B3, B7, B8, B10, B12*	*B1, B2, B3, B4, B5, B6, B7, B8, B9, B10, B11, B12*	*B3, B7, B8, B10, B12*	*I*
B8	*B1, B3, B7, B8, B10, B11, B12*	*B1, B4, B6, B7, B8, B11, B12*	*B1, B7, B8, B11, B12*	*III*
B9	*B2, B3, B7, B9, B11, B12*	*B6, B9*	*B9*	*IV*
B10	*B3, B7, B10, B11, B12*	*B1, B4, B7, B8, B10, B11, B12*	*B7, B10, B11, B12*	*II*
B11	*B2, B3, B7, B8, B10, B11, B12*	*B1, B4, B8, B9, B10, B11*	*B8, B10, B11*	*III*
B12	*B7, B8, B10, B12*	*B1, B2, B4, B5, B6, B7, B8, B9, B10, B11, B12*	*B7, B8, B10, B12*	*I*

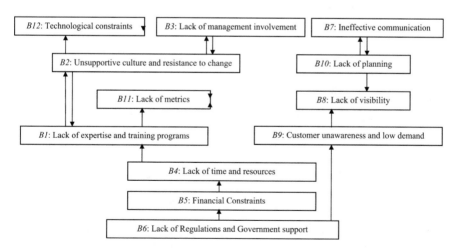

Fig. 2 ISM-based model of barriers in SGRLM

Lack of expertise and training programs (*B1*) may be also caused by the lack of resources, financial constraints and also the lack of government support to change educational programs so they can include SGRLM-related topics. The lack of regulations and government support may also influence the customer awareness (*B9*).

The lack of metrics (*B11*) and the lack of visibility (*B8*) come in the next level. Unsupportive culture and resistance to change (*B2*) and lack of planning (*B10*) from the next level indicating towards ineffective communication (*B7*), lack of management involvement (*B3*) and technological constraints (*B12*) are the top-level barriers in the model.

6   MICMAC Analysis

Cross-impact matrix multiplication applied to classification (MICMAC) aims to examine 'the driving power and dependence power of enablers. This analysis is carried out to distinguish the main enablers (referred to in this paper as barriers) that drive the model in a variety of classes' (Gholami et al., 2020). The driving power and dependence power for each barrier are tabulated in Table 5.

The barriers of SGRLM are categorized into four clusters, as presented in Fig. 3.

- *Quadrant 1* shows the 'autonomous barriers' that have weak dependence and driving powers. Based on Fig. 3, lack of planning *(B10)* does not have much influence on the model.

Table 5 Driving power and dependence power of SGRLM diagram

Barrier	B1	B2	B3	B4	B5	B6	B7	B8	B9	B10	B11	B12	Driving power
B1	1	1	1^a	0	0	0	1	1^a	0	1	1	1	8
B2	1	1	1	0	0	0	1	0	0	0	0	1	5
B3	1	1	1	0	0	0	1	0	0	0	0	0	4
B4	1	1	1	1	0	0	1	1^a	0	1	1^a	1^a	9
B5	1	1	1	1	1	0	1^a	0	0	0	0	1	7
B6	0	1	1	0	1	1	1^a	1^a	1	0	0	1	8
B7	0	0	1^a	0	0	0	1	1	0	1	0	1	5
B8	1^a	0	1	0	0	0	1	1	0	1^a	1	1	7
B9	0	1	1	0	0	0	1^a	0	1	0	1^a	1^a	6
B10	0	0	0	0	0	0	1	0	0	1	1	0	3
B11	0	1^a	1	0	0	0	1	1	0	1^a	1	1^a	7
B12	0	0	0	0	0	0	1	1	0	1^a	0	1	4
Dependence power	6	8	10	2	2	1	12	7	2	7	6	10	

^{a}Entries indicates the presence of transitivity

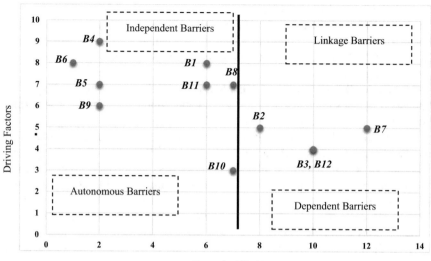

Fig. 3 MICMAC analysis

- *Quadrant 2* shows the 'dependent barriers that have weak driving power but a high dependence power'. Barriers in this category are lack of management involvement *(B3)* and technological constraints *(B12)*.
- *Quadrant 3* shows the 'linkage barrier that have both strong driving and dependence powers'. Unsupportive culture and resistance to change *(B2)* and ineffective communication *(B7)* fell into this quadrant.
- *Quadrant 4* shows the 'independent barriers that have strong driving power but weak dependence power'. As shown in Fig. 3, barriers in this category are lack of expertise and training programs *(B1)*, lack of time and resources *(B4)*, financial constraints *(B5)*, lack of regulations and government support *(B6)*, lack of visibility *(B8)*, customer unawareness and low demand *(B9)* and lack of metrics *(B11)*.

7 Results and Discussion

The objective of this study is to identify and analyse barriers facing the implementation of SGRLM in manufacturing industries. After identifying the most cited barriers, the teamwork of experts was able to provide insights on how these barriers can become relevant and adequate for the new concept of SGRLM, since the barriers of the paradigms constituting these concepts were reviewed one or two at a time.

The lack of regulations and government support was found to be the most significant driver barrier in the implementation of SGRLM. Next, financial constraints have been also analysed as a significant barrier in the shift towards SGRLM. Furthermore, the lack of time and resources and the lack of expertise and training programs are one the crucial barriers hindering the successful launch of SGRLM. These findings can be supported by the literature. Singh et al. (2020) have also found the lack of government support as a significant challenge for green lean practices. Soti et al. (2010) also cite the fund allocation and expert training as important next-level driving enablers.

In order to start the process of SGRLM, government support needs to be present, supported by the sufficient and efficient fund allocation. Overcoming these barriers will pave way for starting the activities of SGRLM.

As a result, the lack of time and resources will be overcome, since funds are now available to recruit additional staff to support the activities of SGRLM. The government support will also result in influencing customers and in raising awareness about the benefits that escort this new mode of manufacturing. The allocation of funds will help get the needed expertise to support SGRLM and to train staff. Also, with the help of government, more emphasis will be put on educational programs and more collaborations with universities.

With the necessary training, staff is now aware of the needed metrics that will monitor the implementation of SGRLM; therefore, more visibility is granted. Also, with the awareness of customers, their demand will become more or less stable which will enhance even more the visibility and improve forecasting. Lack of planning has been found to not have influence on the model; this can be explained by the fact the new technologies of I4.0 will enhance the planning capacity in real time (Pereira et al., 2019). Besides more planning provides more visibility.

Training programs will result in elevating the culture of the organizations, where more awareness will be raised about the relevancy of SGRLM and vice versa and more awareness will push firms to provide more programs. Finally, it has been pointed out as an outcome of the present study that all the barriers converge at technological constraints, lack of management involvement and ineffective communication.

8   Conclusion

In this paper, an attempt has been made to identify and analyse the barriers to implement SGRLM through an extensive literature review. SGRLM is an emerging research area in today's competitive and environmentally conscious world. In the recent past, researchers have investigated the barriers of these paradigms individually; however, no work has been done to investigate the barriers that face the integration of these four paradigms. For all we know, this paper is the first to investigate and analyse the barriers that hinder SGRLM's successful implementation.

Twelve significant barriers to SGRLM process application have been found appropriate to be analysed. These barriers were then analysed by the means of ISM technique. Next the MICMAC analysis enabled us to group these barriers into four categories, which are driving (independent), linkage, dependent and autonomous barriers. Results show that the lack of regulations and government support and financial constraints are the main barriers that face SGRLM.

This study provides multiple managerial contributions. First, it introduces the importance of investigating the barriers faced by SGRLM. The use of ISM-based model helped to identify the impact of managerial efforts made on 'driving variables' and 'linkage variables' identified in the MICMAC analysis. By tackling these barriers, the manufacturing industry will be better prepared to face uncertainty and overcome its vulnerabilities, thus shifting towards operational excellence.

This study has some limitations too, since the analysis is highly qualitative and based on the judgements and opinions of the experts. Furthermore, the barriers used for the analysis come from the manufacturing industry in general and not from a specific industry. However, this can be overcome by using fuzzy and grey set theories or structural equation modelling in specific manufacturing sectors.

References

Amjad, M. S., Rafique, M. Z., Hussain, S., & Khan, M. A. (2020). A new vision of LARG manufacturing—A trail towards industry 4.0. *CIRP Journal of Manufacturing Science and Technology, 31*, 377–393.

Amjad, M. S., Rafique, M. Z., & Khan, M. A. (2021). Modern divulge in production optimization: An implementation framework of LARG manufacturing with industry 4.0. *International Journal of Lean Six Sigma, 12*(5), 992–1016. https://doi.org/10.1108/IJLSS-07-2020-0099

Attri, R., Dev, N., & Sharma, V. (2013). Interpretive structural modelling (ISM) approach: An overview. *Research Journal of Management Sciences, 2319*(2), 1171.

Diekmann, A., & Preisendörfer, P. (2003). Green and greenback. *Rationality and Society, 15*(4), 441–472. https://doi.org/10.1177/1043463103154002

Ghazilla, R. A. R., Sakundarini, N., Abdul-Rashid, S. H., Ayub, N. S., Olugu, E. U., & Musa, S. N. (2015). Drivers and barriers analysis for green manufacturing practices in Malaysian Smes: A preliminary findings. *Procedia CIRP, 26*, 658–663. https://doi.org/10.1016/j.procir.2015.02.085

Gholami, H., Bachok, M. F., Saman, M. Z. M., Streimikiene, D., Sharif, S., & Zakuan, N. (2020). An ISM approach for the barrier analysis in implementing green campus operations: Towards higher education sustainability. *Sustainability, 12*(1), 363.

Govindan, K., Diabat, A., & Shankar, K. M. (2015). Analyzing the drivers of green manufacturing with fuzzy approach. *Journal of Cleaner Production, 96*, 182–193.

Heinicke, M. (2014). *Framework for resilient production systems*. In IFIP international conference on advances in production management systems (pp. 200–207). Springer.

Jharkharia, S., & Shankar, R. (2005). IT-enablement of supply chains: Understanding the barriers. *Journal of Enterprise Information Management, 18(1)*, 11–27. https://doi.org/10.1108/17410390510571466

Jones, D. A., Wang, W., & Fawcett, R. (2009). High-quality spatial climate data-sets for Australia. *Australian Meteorological and Oceanographic Journal, 58*(4), 233.

Kamble, S. S., Gunasekaran, A., & Sharma, R. (2018). Analysis of the driving and dependence power of barriers to adopt industry 4.0 in Indian manufacturing industry. *Computers in Industry, 101*(June), 107–119. https://doi.org/10.1016/j.compind.2018.06.004

Kang, H. S., Lee, J. Y., Choi, S. S., Kim, H., Park, J. H., Son, J. Y., Kim, B. H., & Do Noh, S. (2016). Smart manufacturing: Past research, present findings, and future directions. *International Journal of Precision Engineering and Manufacturing-Green Technology, 3*(1), 111–128.

Klein, M. M., Biehl, S. S., & Friedli, T. (2018). Barriers to smart services for manufacturing companies—An exploratory study in the capital goods industry. *Journal of Business and Industrial Marketing, 33*(6), 846–856. https://doi.org/10.1108/JBIM-10-2015-0204

Kumar, S., Luthra, S., Govindan, K., Kumar, N., & Haleem, A. (2016). Barriers in green lean Six Sigma product development process: An ISM approach. *Production Planning & Control, 27*(7–8), 604–620. https://doi.org/10.1080/09537287.2016.1165307

Kusiak, A. (2018). Smart manufacturing. *International Journal of Production Research, 56*(1–2), 508–517.

Kusiak, A. (2020). *Resilient manufacturing*. Springer.

Lee, H.-T., Song, J.-H., Min, S.-H., Lee, H.-S., Song, K. Y., Chu, C. N., & Ahn, S.-H. (2019). Research trends in sustainable manufacturing: A review and future perspective based on research databases. *International Journal of Precision Engineering and Manufacturing-Green Technology, 6*(4), 809–819.

Lohmer, J., & Lasch, R. (2020). Blockchain in operations management and manufacturing: Potential and barriers. *Computers and Industrial Engineering, 149*(March), 106789. https://doi.org/10.1016/j.cie.2020.106789

Matos, F., & Jacinto, C. (2019). Additive manufacturing technology: Mapping social impacts. *Journal of Manufacturing Technology Management, 30*(1), 70–97. https://doi.org/10.1108/JMTM-12-2017-0263

Melnyk, S. A., & Smith, R. T. (1996). *Green manufacturing*. Computer Automated Systems of the Society of Manufacturing Engineers.

Melton, T. (2005). The benefits of lean manufacturing: What lean thinking has to offer the process industries. *Chemical Engineering Research and Design, 83*(6), 662–673.

Mittal, V. K., & Sangwan, K. S. (2014). Prioritizing barriers to green manufacturing: Environmental, social and economic perspectives. *Procedia CIRP, 17*, 559–564. https://doi.org/10.1016/j.procir.2014.01.075

Mittal, V. K., Sindhwani, R., & Kapur, P. K. (2016). Two-way assessment of barriers to lean–Green manufacturing system: Insights from India. *International Journal of Systems Assurance Engineering and Management, 7*(4), 400–407. https://doi.org/10.1007/s13198-016-0461-z

Mollenkopf, D., Stolze, H., Tate, W. L., & Ueltschy, M. (2010). Green, lean, and global supply chains. *International Journal of Physical Distribution and Logistics Management, 40*(1–2), 14–41. https://doi.org/10.1108/09600031011018028

Mudgal, R. K., Shankar, R., Parvaiz, T., & Tilak, R. (2010). Modelling the barriers of green supply chain practices. *International Journal Logistics Systems and Management, 7*(1), 81–107. http://www.inderscience.com/link.php?id=13460

Paul, I. D., Bhole, G. P., & Chaudhari, J. R. (2014). A review on green manufacturing: It's important, methodology and its application. *Procedia Materials Science, 6*, 1644–1649.

Pavnaskar, S. J., Gershenson, J. K., & Jambekar, A. B. (2003). Classification scheme for lean manufacturing tools. *International Journal of Production Research, 41*(13), 3075–3090.

Peillon, S., & Dubruc, N. (2019). Barriers to digital servitization in French manufacturing SMEs. *Procedia CIRP, 83*, 146–150. https://doi.org/10.1016/j.procir.2019.04.008

Pereira, A. C., Dinis-Carvalho, J., Alves, A. C., & Arezes, P. (2019). How industry 4.0 can enhance lean practices. *FME Transactions, 47*(4), 810–822.

Pettit, T. J., Croxton, K. L., & Fiksel, J. (2019). The evolution of resilience in supply chain management: A retrospective on ensuring supply chain resilience. *Journal of Business Logistics, 40*(1), 56–65.

Rajesh, R. (2018). Measuring the barriers to resilience in manufacturing supply chains using grey clustering and VIKOR approaches. *Measurement: Journal of the International Measurement Confederation, 126*(May), 259–273. https://doi.org/10.1016/j.measurement.2018.05.043

Rimanoczy, I., & Pearson, T. (2010). Role of HR in the new world of sustainability. In *Industrial and commercial training*. Emerald Group Publishing.

Ruben, B., Vinodh, S., & Asokan, P. (2016). ISM and fuzzy MICMAC application for analysis of lean six sigma barriers with environmental considerations. *International Journal of Lean Six Sigma, 9*(1), 64–90. https://doi.org/10.1108/IJLSS-11-2016-0071

Shah, R., & Ward, P. T. (2003). Lean manufacturing: Context, practice bundles, and performance. *Journal of Operations Management, 21*(2), 129–149.

Shibin, K. T., Gunasekaran, A., Papadopoulos, T., Dubey, R., Singh, M., & Wamba, S. F. (2016). Enablers and barriers of flexible green supply chain management: A total interpretive structural modeling approach. *Global Journal of Flexible Systems Management, 17*(2), 171–188. https://doi.org/10.1007/s40171-015-0109-x

Sindhwani, R., Mittal, V. K., Singh, P. L., Aggarwal, A., & Gautam, N. (2019). Modelling and analysis of barriers affecting the implementation of Lean Green Agile Manufacturing System (LGAMS). *Benchmarking, 26*(2), 498–529. https://doi.org/10.1108/BIJ-09-2017-0245

Singh, C., Singh, D., & Khamba, J. S. (2020). Analyzing barriers of green lean practices in manufacturing industries by DEMATEL approach. *Journal of Manufacturing Technology Management, 32*(1), 176–198. https://doi.org/10.1108/JMTM-02-2020-0053

Soti, A., Shankar, R., & Kaushal, O. P. (2010). Modeling the enablers of six sigma using interpreting structural modeling. *Journal of Modelling in Management, 5*(2), 124–141.

Sundar, R., Balaji, A. N., & Satheesh Kumar, R. M. (2014). A review on lean manufacturing implementation techniques. *Procedia Engineering, 97*, 1875–1885.

Tao, F., Qi, Q., Liu, A., & Kusiak, A. (2018). Data-driven smart manufacturing. *Journal of Manufacturing Systems, 48*(July), 157–169. https://doi.org/10.1016/j.jmsy.2018.01.006

Webel, S., Bockholt, U., Engelke, T., Gavish, N., Olbrich, M., & Preusche, C. (2013). An augmented reality training platform for assembly and maintenance skills. *Robotics and Autonomous Systems, 61*(4), 398–403. https://doi.org/10.1016/j.robot.2012.09.013

Womack, J. P., & Jones, D. T. (1996). Beyond Toyota: How to root out waste and pursue perfection. *Harvard Business Review, 74*(5), 140–172.

Yu, C., Xun, X., & Yuqian, L. (2015). Computer-integrated manufacturing, cyber-physical systems and cloud manufacturing—Concepts and relationships. *Manufacturing Letters, 6*, 5–9.

Zhang, A., Venkatesh, V. G., Liu, Y., Wan, M., Ting, Q., & Huisingh, D. (2019). Barriers to smart waste management for a circular economy in China. *Journal of Cleaner Production, 240*. https://doi.org/10.1016/j.jclepro.2019.118198

Zheng, P., Wang, H., Sang, Z., Zhong, R. Y., Liu, Y., Liu, C., Mubarok, K., Yu, S., & Xu, X. (2018). Smart manufacturing systems for industry 4.0: Conceptual framework, scenarios, and future perspectives. *Frontiers of Mechanical Engineering, 13*(2), 137–150. https://doi.org/10.1007/s11465-018-0499-5

Secure Model for Records Traceability in Airline Supply Chain Based on Blockchain and Machine Learning

Karim Zkik, Anass Sebbar, Narjisse Nejjari, Sara Lahlou, Oumaima Fadi, and Mustapha Oudani

1 Introduction

The airline industry is certainly one of the fastest growing industries of the twenty-first century. It is getting more complex due to the diversity of intermediate transactions between suppliers, contractors, employees, and end customers. Blockchain technology, when applied to the supply chain, could provide efficient ways to deal with this complexity, in terms of data storage, data management, traceability, and security (Zheng, 2018). According to Market and Market (Mehra, 2020), the airline blockchain market is expected to grow from USD 421 million in 2019 to USD 1.394 million by 2025. Furthermore, the major specificity of the use of blockchain in the supply chain is that blockchain provides a much higher level of security in terms of data integrity and privacy. Yet, the most significant promise of blockchain for the supply chain is improved data security for digital records used throughout all the supply chain process (Queiroz, 2019). Thus, because of its distributed nature, blockchain is well suited to validating the authenticity of hardware and software, which are critical in terms of airline supply chain security management.

The deployment of blockchain technology in the airline supply chain requires a focus on blockchain challenges (Tasatanattakool, 2018). Data security and transparency have recently emerged as critical concerns for airline industries, as big data has

K. Zkik (✉)
ESAIP Graduate School of Engineering, CERADE, Angers, France
e-mail: kzkik@esaip.org

A. Sebbar · N. Nejjari
International University of Rabat, TICLAB, Rabat, Morocco

ENSIAS Rabat, Mohammed V University, ESIN, Rabat, Morocco

S. Lahlou · O. Fadi · M. Oudani
International University of Rabat, TICLAB, Rabat, Morocco

grown at an exponential rate. Since the centralized management of the supply chain is putting the data integrity and availability at multiple risks, many works have introduced a blockchain architecture that ensures the traceability of the airline supply chain as well as promotes the data flow management within the supply chain (Di Vaio, 2020; Santonino III, 2018). However, most of these models are more concerned with data confidentiality and integrity rather than availability, resiliency, and non-repudiation. Consequently, there is still a need for more effective and practical solutions that better optimizes the effectiveness of BT in the ASCM realm.

The main goal of this work is to build a framework that deals with data traceability and transaction security in ASCM using blockchain and machine learning techniques. This framework allows authenticating legitimate users who participate in transactions in the airline supply chain and to validate transactions through a smart application based on smart contracts. Furthermore, to strengthen the security of our network and to mitigate advanced and persistent attacks, we propose a security module based on machine learning for anomaly detection and preservation. Indeed, machine learning as the core of artificial intelligence allows us to promote the privacy of the supply chain and make it smarter and more secure.

The remainder of the paper proceeds as follows. We start by studying the related works that integrated the blockchain in the airline supply chain in Sect. 2; in Sect. 3 we propose a secure model for traceability records and security in airline supply chain based on blockchain and machine learning. In Sect. 4, we discuss our findings including the efficiency of our proposed model. We present limitations, future work, and conclude in the last section.

2   Related Work

(a) *Introduction to Airline Supply Chain Management.*

A supply chain management or (SCM) is a set of steps that lead to the delivery of a product or service to a client (Flynn, 2018). In other words, a supply chain is a network that connects a business with its vendors to manufacture and deliver a particular commodity to the end user. Moving and converting raw materials into finished goods, shipping those products, and selling them to end users are all stages in the process (Khan, 2021). Overall, supply chain management allows businesses to increase their profit margins in a variety of ways, and it is particularly important for businesses with large and multinational operations.

In the airline industry realm, supply chain management (SCM) is widely used as a high strategic practice to increase the number of aircraft systems that need to be maintained and repaired by following the growth and the development of new aircraft technologies (Alshurideh, 2019). Figure 1 presents the standard supply chain of airline workflow with maintenance support and the main actors in aircraft supply chain management (ASCM).

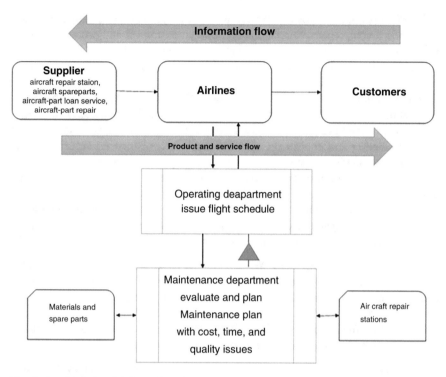

Fig. 1 Supply chain of airline with maintenance support

In this scenario, we have several types of suppliers who offer a wide range of products and maintenance services to airlines such as aircraft repair stations, aircraft spare, aircraft part loans, and aircraft part repair services. Then we have the airline firms that provide services to passengers, travel agencies, and other airlines while respecting procedures, rules, and deadlines.

On the other hand, the airline industry must maintain effective and efficient maintenance strategies (Ertogral, 2019). To do so, the back office must plan future maintenance in line with the flight plan. Furthermore, managers must decide in advance whether to outsource or outsource maintenance services and material suppliers. These activities require powerful and influential decision-making tools for aircraft maintenance and related supply chains. In addition, the planner must investigate the flight requirements of the aircraft and convert them into a flight plan that accurately indicates the registration number and flight schedule of the aircraft. Thus, to provide an effective supply chain in the airline context and prevent costly bottlenecks, several studies have been proposed in several axes such as planning, procurement of raw materials, processing, and distribution. Ball et al. (Ball, 2007) proposed a commercial air transportation value chain that is made up of many interconnected parts. They split into upstream and downstream branches, with airlines serving as the central node. On the other hand, Westerkamp and Friedhelm

(Westerkamp, 2018) proposed a decentralized SCM system based on smart contracts to provide a comprehensive relationship between customers and producers during the manufacturing processes to provide information in a timely and reliable manner.

Despite all these innovative solutions, airline supply chains are becoming more complex due to the interplay of many actors than before and the structurally disjointed entities competing to serve customers (Bode, 2015) Among the challenges that can be faced by ASCM regarding Kshetri's research (Kshetri, 2017) are lack of visibility from the upstream side to the downstream side, lack of flexibility to sudden demand changes and control on operation cost, lack of trust on security among stakeholders, ineffective supply chain risk management, and lack of advanced technologies. Furthermore, airline industry is increasingly evolving and duplicating much effort in the quest for strategic and administrative strategies to achieve long-term success while avoiding the negative consequences of technological disruptions. Thus, as airline companies expand, there is a need to smartly face challenges related to the management of the supply chain such as:

Smart Inventory Management When an airline company has too many suppliers, the problem of inventory management can cause severe delays or an excess of aircraft parts, which can be costly (Bodkhe, 2020). Companies must identify a course of action to balance their inventory holdings efficiently through predictive analysis. Thus, the importance of real-time data, as well as data storage and manipulation provided by blockchain technologies, cannot be overstated.

The Connection Between the Company, Suppliers, and Customers Customer expectations are increasing as more competitors are present in the market, providing customers with more options. Now minimal flight disruptions are expected. To meet high customer expectations, companies must keep a close eye on both customers and competitors. Being better and more connected with suppliers and customers will help build trust and loyalty from both. Effective collaboration is primordial.

The Use of More Advanced Technologies The cadence of technological change is growing rapidly challenging the ability of the aircraft industry to keep up with the flow (Rejeb, 2021). The challenge here is to identify which changes offer the optimal value to the industry while considering the quality and cost of products and services provided.

Supply Chain Optimization Maintenance capacity, availability of labor, physical location around the world, and the quickest possible reaction time without overcommitting extremely valuable capital investment are also part of challenges (Sawik, 2020).

The objective behind this work is to provide maintenance repair and overhaul with the optimal solutions to face challenges and optimize the airline supply chain. When the system's cost and complexity rise, optimization provides more value to the customers.

(b) *Blockchain for Supply Chain Management.*

Supply chain management covers several vertical levels, hundreds of horizontal partnerships, multiple geographic areas, different financial processes and payment mechanisms, and different temporal stresses. However, these measurements are difficult to manage while applying continuous improvement strategies (Kshetri, 2017; Belhad, 2020). Several studies confirm that emerging technology is the key to meet improvement targets in the supply chain context (Belhadi, 2021). In fact, supply chain is continuously integrating new technologies, such as big data, IoT, Industry 4.0, and blockchain technologies (BT). Blockchain is a new technology that permits to prove the authenticity of exchanged data and to authenticate nodes participating in communications in decentralized environments (Westerkamp, 2018). The main characteristics of the blockchain are the identification of nodes, the validation of transactions through a consensus mechanism, and the decentralization of communication security management. The use of the blockchain makes it possible to promote transparency, guard against the risks of embezzlement, and gain in productivity and efficiency. The blockchain network is made up of several nodes that keep track of a collection of shared data and carry out transactions that alter the data. Until being ordered and packed into a timestamped block, transactions must be authenticated by most network nodes. This mining method is dependent on the blockchain network's consensus framework. Then, both network nodes check that the current suggested block includes legitimate transactions and references the right previous block with a cryptographic pointer before connecting it to the list. Thus, BT permits to keep track of all variations in decentralized architectures to secure and to validate the authenticity of transactions without the need for a third party like governments or banks (Bode, 2015).

BT has also been introduced supply chain field to make the chain more transparent, trustworthy, and authentic (Kamble, 2021). In fact, BT has the potential to handle various security issues as it can eliminate the need for the centralized authority to perform various operations (Francisco, 2018). From January to June 2016, yellowfin and skipjack tuna fish were tracked throughout the entire supply chain, from fishermen to distributors over a traceability project enabled by Ethereum (Earley, 2013). End users could then track the "story" of their tuna fish sandwiches via a smartphone and determine information about the producers, suppliers, and procedures undergone by the end product. Everledger (Azzi, 2019) is another blockchain enabled traceability application for the global diamond industry. The company, which partnered with Barclays, created a database of over a million diamonds registered on their blockchain to certify the final cut diamond was ethically sourced from "conflict-free" regions.

There are numerous works that integrated blockchain technology to manage the supply chain. For example, Ambrosus (Kirejczyk, 2017) was proposed as a public blockchain ecosystem that aims to promote the traceability of the transmitted products; it uses sensors and tags to follow up the products through the supply chain process; the sensors affected to the product record the data (e. g., temperature and humidity). So, the integrity of the shipment will be ensured since any compromised behavior will be detected. Modum (Bocek, 2017) is also a public blockchain system that monitors and traces products in a way to comply with the GDP (good distribution practice) regulations. Modum mainly aims to reduce the number of intermediaries in the logistics process as well as to monitor the products by IoT devices and sensors. The integration of blockchain in this solution makes the supply chain tamper-proof, and by that, it will ensure that the data is immutable and shared between all the supply chain members securely.

Problem Statement

The review of the literature shows that there are several works that deal with BT as an instrument that ensures traceability and privacy for data management system for the supply chain. However, these solutions have not been set sufficiently into the context of the improvement of airline operations using supply chain management through blockchain technology. Thus, there is still a need for more effective and practical solutions that better optimizes the effectiveness of BT in the ASCM realm. Furthermore, the governance and security of the airline system are known to be very critical. Subsequently, the gap of vulnerabilities needs to be filled. In fact, studies that focus on the airline's sustainable supply chain coupled with blockchain are scarce. Indeed, Madhwal et al. highlighted this issue and show up its importance for enhancing airline's supply chain transparency (Madhwal, 2017). In addition, existing studies in literature are more focused on managerial aspects than effective solutions with real simulations.

On the other hand, BT still suffers from many challenges related to governance and security issues (Appelhanz, 2016). According to several research studies, blockchain technologies are still vulnerable to several types of attacks such as hacks exchange, malware infections, software flaws, and 51% attacks. Several security models have been proposed in the literature to secure blockchain technologies using anonymous signatures, homomorphic encryption (HE), and attribute-based encryption (ABE). These solutions are mainly based on data encryption. However, these mechanisms are very greedy in terms of calculation and storage resources. Furthermore, most of these models are more focused on data confidentiality and integrity without considering availability, resiliency, and non-repudiation.

Machine learning (ML) can be applied to blockchains based applications to make them smarter and more secure. In fact, ML enables to build better models for different applications, such as fraud detection and identity theft detection (Sebbar, 2020). Several research works in the literature provide different approaches to improve the security of blockchain networks. For example, Sudeep Tanwar et al. (Bodkhe, 2020) studied how BT and ML can be applied in several smart applications and reviewed the proposed model using machine learning in blockchain-based smart applications. However, despite the interesting results from these studies, to the best of our knowledge, these research works are in their early stages and do not deal with infrastructure availability, quantum resilience, and privacy issues especially in the context of ASCM.

The goal of this project is to find effective solutions for traceability, securing transactions, resilience, and securing maintenance records for airline's supply chain using blockchain and machine learning techniques for ASCM. The research objectives of our project are identified as follows:

- RO1: Provide a reliable security mechanism for traceability records in airline supply chain using BT.
- RO2: Identify the security issues in the realm of blockchain-based ASCM.
- RO3: Investigate the use of machine learning techniques on blockchain-based ASCM architectures.

3 Proposed Secure Model for Records Traceability in Airline Supply Chain Based on Blockchain

The main goal of this paper is to provide a new and optimal solution for ASCM to face the previously presented challenges. Blockchains have received, lately, much interest worldwide. It became used in different areas including cryptocurrency, healthcare, energy, insurance, copyright protection, IoT, advertising, and societal applications. Moreover, the airline industry is one of the most areas in need of the implementation of blockchain technology since it is ruled by safety concerns and heavy regulations. An implementation that will help aerospace companies reduce maintenance costs, increase aircraft availability, and minimize errors in tracking aircraft parts because this elegant and paradigm-shifting technology has the potential to deliver profound benefits for the hundreds of suppliers typically involved in the manufacturing of a single aircraft.

In this section, we propose a framework that secures blockchain-based smart applications in the aircraft supply chain. This section will be organized as follows; the first subsection highlights the authentication phase in our proposed blockchain model. The second is about how the process of consensus is ensured in our model. The third part concerns the integration of machine learning in our blockchain model, specifically while using anomaly detection systems to provide more security for the proposed system.

Before defining the authentication process, we should first highlight the components of our model. The corresponding system architecture is composed of four layers as shown in Fig. 2.

The first layer is responsible of collecting data using IoT sensors that have been specifically deployed for this purpose. GPS, for example, is used to locate products in the supply chain, and RFID technology is used to record information such as quality and transaction information. The second layer concerns data storage. Data is stored in four distributed ledgers: quality data, logistics data, assets data, and transaction data. The main goal of this layer is to facilitate data sharing and assists quality control in the supply chain. On the other hand, privacy issues can be the first concern; that's why many mechanisms are used in the contract layer, such as digital

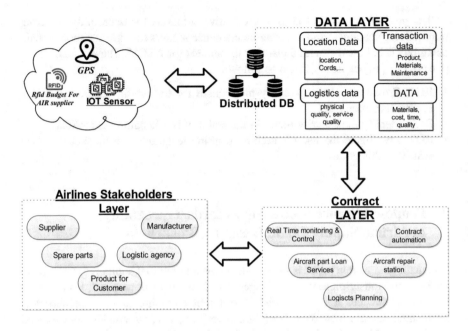

Fig. 2 Components of the blockchain-based system

identity to control the access authority to the data, smart contract automation to ensure data transmission between the supply chain nodes. Finally, airline stakeholders layer includes all business actors who control and manage our blockchain model through decision-making, purchasing, and manufacturing activities.

3.1 Authentication Process for ASCM System

The airline supply chain deals with crucial data and the blockchain model is deployed to ensure the security of the users, customers, and equipment's data. Moreover, it is a high priority to control the access of users and customers to the airline supply chain as well as to assure their identity. Thus, in our approach, a multi-authentication process is used to ensure transaction with the end user. The multi-authentication approach used adds a next level of security to the blockchain network and meets the regulatory compliance requirements of the supply chain. Figure 3 illustrates the sequence diagram of the process of multi-authentication.

Our solution is structured into front-end, back-end, and different types of tags (RFID, GPS, CA) that will ensure the users/equipment's authentication. As outlined in Fig. 3, our architecture contains the following components:

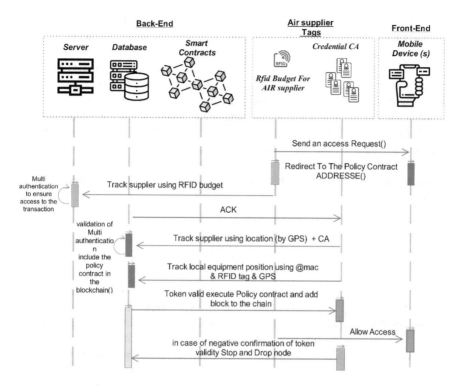

Fig. 3 Multi-authentication process to ensure transaction with clients for ASCM

- *Front-end*: It embeds mobile devices used by the end users to send authentication request and transmit data to other parties via blockchain network.
- *Air supplier tags*: Such as RFID (radio frequency ID), GPS location, and confidential CA. All those identifiers are collected from the deployed sensors and used to ensure an efficient authentication process.
- *Back-end*: Output level of our solution which is responsible for producing and displaying the final results and it contains the following:

  – *Smart contracts*: Are issued for integrating new members and equipments after their authentication.
  – *Server*: In which the RFIDs, GPS, and credential CA data are processed, stocked, and verified, also in which the smart contracts are executed and deployed.
  – *Database*: It is a relational database used to store the users/equipment's data and maps between the end users' transactions and their smart contracts.

The main goal of the multi-authentication process is to securely authenticate legitimate users and to ensure transaction's validity with end clients. Thus, to start the authentication process, a supplier sends an authentication request to access the application. Through the front end, an RFID must be provided to track the user; the

RFID is then verified within the back end. After the first verification, an acknowledgment must be returned to the credential CA. Other layers of authentication are applied to validate the user access by tracking users using location (GPS) and equipment's using their Mac address. Thereafter, if the token validates the user/equipment request, the block is added to the blockchain and the access is validated, if not a negative confirmation is declared, and the node is dropped.

After the deployment of a multi-authentication system, the airline supply chain is now in full control of who is allowed to access data, join the blockchain network, and launch transactions. After the authentication phase, the eligible end users are now allowed to join the blockchain network and launch their transactions. For that matter, we must specify the consensus mechanism appropriate to manage the new network members.

3.2 Consensus Process for ASCM System

To manage network transactions and ensure security when receiving and responding to requests from highly sensitive sensors in ASCM environment, blockchain is used in our approach. Indeed, the blockchain can offer flexibility to apply various transaction rules to ensure and facilitate the management of airlines. Thus, the proposed blockchain framework aims to ensure the basic benefits of blockchain technology such as accessibility, visibility, transparency, and traceability. The global architecture of our model is structured into back-end, front-end, and based on the use of GPS sensors from mobile devices.

In our current implementation, creating a new block with modified transactions, calculating the hash, and replacing it with an older block are not a problem. Nonetheless, we need a way to ensure that any changes from previous blocks invalidate the entire chain. For this, the airline supplier chain builds a dependency between consecutive blocks by chaining them with the block hash immediately preceding them. Therefore, through chaining, we embed the hash from the previous block into the current block in a new field called the previous hash. The genesis block can be generated manually or through unique logic. To do this, we will use the previous_hash field to implement the initial structure of our blockchain class. To attach a block to the chain, it will first be necessary to verify that the details have not been manipulated (the proof of the work provided is correct) and that the sequence of transactions (the previous hash field of the block to be inserted points to the hash of the last block in our chain) is kept. The proposed consensus phase is composed of five stages:

1. *Proof of work (PoW)*: This phase concerns the implementation of the proof of work algorithm.
2. *Block generation*: In this phase we add needed blocks to the chain. Thus, a distributed data structure is used to store and list all transactions.

3. *Block chaining*: This phase permits to chain a sequence of blocks in a specific order to facilitate the storage and to prevent attacks such as 51% attacks.
4. *Mining phase*: In this phase we will select some specific nodes to perform the verification process. These nodes are considered as miners in the network.
5. *Consensus phase*: A set of rules and arrangements to carry out blockchain operations are performed in this phase.

Algorithm 1: Consensus Phases
Input: Block, proof, block_hash, self.
Output: Validity of Consensus *// a set of rules and arrangements to carry out blockchain operations.*

    function PoW(*self, block*).
        block.nonce←0
        computed_hash ← block.compute_hash()
        while (compute_hach(*'00'*blockchain*)! = 0) do.
            block.nonce + = 1.
            computed_hash ← block.compute_hash()
        return computed_hash.
    function add_block(*self, block, proof*).
        if (previous_hash! = block.previous_hash) then
            return False.
        else Blockchain.is_valid_proof(*block, proof*)
            return False.
        block.hash ← proof.
        self.chain.append(*block*).
        return True.
    function proof_validity(*self, block, block_hash*).
        return (block_hash.startswith(*'00' * blockchain*) && block_hash == block.
compute_hash()).
    function add_new_transaction(*self, transaction*).
        self.unverified_transactions.append(*transaction*).
    function mine(self):
        if (not self.unverified_transactions).
            return False.
        last_block←self.last_block.
        new_block ← Block(*index* ← *last_block.index* + *1, transactions* ← unverified_*transactions,*
        previous_hash← *last_block.hash*).
        proof ← self.proof_of_work(new_block).
        self.add_block(new_block, proof).
        self.unverified_transactions ← [].
    return new_block.index.

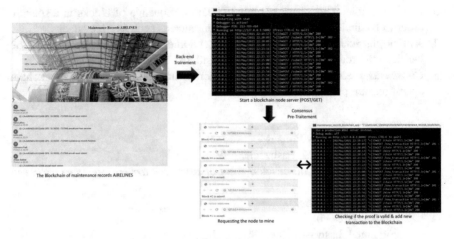

Fig. 4 ASCM application – front-end and back-end screens

As shown in Algorithm 1, the transactions will be initially stored as a pool of unconfirmed transactions. The process of putting the unconfirmed transactions in a block and computing proof of work is known as the mining of blocks. Once the nonce satisfying our constraints is figured out, we can say that a block has been mined and it can be put into the blockchain. The proposed system is based on the deployment of smart contracts, as shown in Fig. 4. These smart contracts are configured and deployed on the server side to ensure overall compliance requirements for each new shipment or group of airline supplier products.

On the other hand, to track transactions efficiently in ASCM, we need to develop an intelligent system for recording GPS data of the location of materials recorded by preinstalled sensors. The GPS data recording mechanism is established as follows:

- *Data request*: In the front end, the web browser and mobile providers communicate with the server using the JSON-based representational state transfer (REST) API. This phase makes it possible to encode and decode the requests and responses of spatial coordinates.
- *Data publication and PoW*: In this phase the system ensures the obtaining and publication of GPS information. In this phase, the validity of credentials and proof of work PoW.
- *Data sharing*: The sender and receiver should then be informed of the outcome of the contract and be able to access the position, preferably using a notification or alert. Thus, the API allows after verification to the recipient the sending and receiving of the GPS coordinates of the location of the materials recorded by the sensors.

3.3 Security Modules for Anomaly Detection and Securing Transactions

The integration of ML models could enhance the level of security in our blockchain model. Machine learning could be used in the detection of abnormal behavior. Abnormal behaviors include abnormal authentication and abnormal transactions in smart contracts.

The main objective is to build a machine learning anomaly-based intrusion detection model that can protect smart contracts compatible with the blockchain we proposed previously. The basic idea behind the proposed model is that the IDS will rollback all the changes to the contract states and raise an alarm to the administrators when an abnormal behavior is detected. The idea is to incorporate the IDS into the smart contracts so that the protection system becomes part of the consensus. The advantage of the anomaly-based approach is that it does not require knowledge of the signatures of known vulnerabilities; instead, it monitors the control flow path directly to classify smart contract execution behavior.

In our case, a predictive ML model will be used to predict the yet-undefined transactions and classify them into normal or malicious transactions. In this regard, the integration of ML models is done in the level of the security module. Thus, we propose a machine learning anomaly-based intrusion detection model that will allow us to enhance the security of our architecture. This mechanism model consists of four stages as shown in Fig. 5.

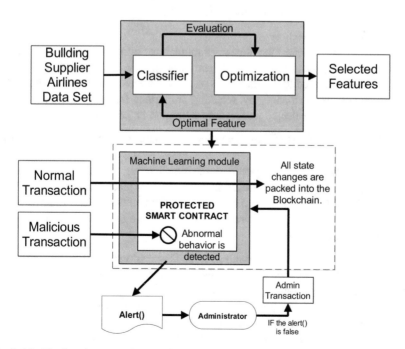

Fig. 5 Machine learning anomaly detection-based blockchain mechanism

Table 1 Important attributes of ASCM

Feature	Description
In-transaction	Number of transactions received per incoming transaction
Out-transaction	Outgoing frequency sent for the supplier per outgoing transaction
Time-in	Average time interval between in-transactions
Time-out	Average time interval between out-transactions
Active-duration	Difference time between first and most recent transactions associated
GPS cords	Cords information for the first supplier associated with a given transaction
Blocks	Number of blocks explored
Transparency and visibility	The transactions or records cannot be hidden

Stage 1: Dataset Acquisition and PreProcessing

In this research, we use the raw Bitcoin blockchain dataset, which (Zambre, 2013) describes the data in the Bitcoin blockchain into a public ledger and is called Bitcoin (BTC) expressed in currency units. The ledger data contains all Bitcoin transactions from the beginning of the network to the present. As recorded by Blockchain.com 2021, there are approximately 4.5 million transactions in the Bitcoin ledger. For each transaction, there can be multiple senders and receiver addresses. Moreover, a single user can have multiple addresses, and each user is anonymous because there is no personal information associated with any address. In order to classify abnormalities transactions from normal, a tree structure is used to represent recursive partitions by training and testing based on random forest model.

Stage 2: Features Selection

In this part, we aim to select the most relevant features from our dataset to lighten the calculations while obtaining high efficiency scores. For this, the feature selection techniques were applied to this dataset to reduce the number of features, without loss of classification performance, by selecting important features to enhance better accuracy results. The selected features are presented in Table 1.

Stage 3: Supervised Machine Learning Anomaly Detection Using Random Forest

In this section, we present a supervised machine learning anomaly detection mechanism using random forest for ASCM framework. This mechanism aims to enhance the work of the blockchain system by automatically identifying and filtering abnormal activities. The target components of the blockchain architecture are the blocks and transactions of a given task.

Random forest is based on classification and regression trees (CART), which is a nonparametric ML method. What makes it flexible and powerful is its ability to detect interesting inter-dependencies between features by using multiple features more than once in different parts of the model. In random forests, tree predictors are combined in such a way that each tree is related to the values of a random vector sampled independently and with similar distribution for all trees which means that

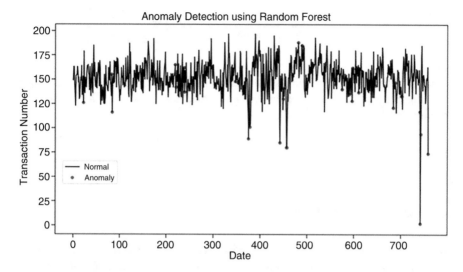

Fig. 6 Anomaly detection rate using random forest

each model receives equal weight. The strength of the individual trees and the correlation between them affect the generalization error of the tree classifier forest.

In this phase we will evaluate our ASCM API by using supervised anomaly detection based on random forest to analyze collected sample. Our model is based on the transaction security policy defined to efficiently detect anomalies. To do so, we divided our dataset into two subsets: training (70%) and testing (30%). The train data is used to fit the machine learning model. Moreover, to show the feasibility of our framework, we will model transaction plots for a better understanding of anomalies classification based on several metrics, such as classification repot, accuracy, and recall, by visualizing them with outlier anomaly. The results of our training phase to detect abnormal transactions are shown in Fig. 6 as a time series diagram.

Abnormal transactions are represented by red dots, and normal transactions are represented by black solid curves per time. The implementation of our model requires a systematic approach to automatically select and compare valid transactions. To determine whether the customized approach is valid, we analyzed the important features (transaction properties) for each individual address. This is to check whether there is a consensus among the important transaction properties to identify anomalies or if they are specific to each address. If the importance factors of the features are consistent across all addresses, there is a clear transaction pattern among all addresses; otherwise, the customized approach is required.

Stage 4: Classification Report Generation

In this stage, we build a classification report to evaluate the performance of the RF classifier model. For each model, we use 70% of the learning process to classify normal data as abnormal based on the threshold of the labeling process of transaction

Table 2 Classification report

Model	Precision	Recall	F1-score	Accuracy
Random forest	0.96	0.80	0.87	0.97

security policy (binary labeling). The remaining datasets are devoted to testing the performance models. Table 2 shows the performance indicators for the training and testing phases.

The obtained results show higher precision, recall, accuracy, and f1-score for the RF classifier.

4 Discussion and Implications

Blockchain technology could be an efficient solution for risk-free method of managing and sharing data about suppliers, material suppliers, and maintenance services in airline companies. In our field, airline records that could be stored on the blockchain are the age of the aircraft, flight inspection date, last aircraft service and repair, changed components, manufacturer of the aircraft parts, and component repair history. The storage of these electronic records in multiple databases could be very challenging. This process could be time-consuming and could lack transparency. In the blockchain, a shared digital journal records all types of transactions, using cryptography to allow all users to securely add new information without the need for a "central" authority. When applied to supply chains, blockchain technology provides the opportunity to secure supply chains while also gaining end-to-end visibility and transparency of physical flows.

The blockchain solution proposed in this work aims to provide improvement to supply chain security related to the airline industry, by exploring data provenance and storage, attribution, and audibility. Smart contracts will provide data visibility by enforcing rules that govern which information each stakeholder has access to. As a basic benefit, the proposed solution provides the following:

- *Accessibility and visibility*: Supply chain data management improves and expedites inter-vendor cooperation.
- *Transparency*: The ability to audit the system and network throughout all its lifecycle.
- *Traceability*: Ensure efficient and secure processes of supply chain components.
- *Monitoring*: Enables auditing and security of critical cyber assets such as software, hardware, and firmware records in a distributed ledger.

On the other hand, all these advantages are dependent on the level of security that the system can provide:

- *Performing multifactor authentication*: Control the execution of blockchain smart contracts using multi-authentication of user and node identities during PoW treatment and validation phases.
- *Auditing behavior events*: The administrator monitors the system in real time to detect abnormal violations and obtain evidence. This could be accomplished by verifying smart contract execution by mining nodes and recording all system acts in accordance with certain norms.
- *Record storage and maintenance*: In our case, all airline supply chain records such as manufacturing data, repair, and maintenance information are stored in a database that is linked to an HTTPS server. In the smart contract, this data includes the date, executor, input, and output information. The blockchain network as a whole must be secured. We propose the use of IDS-IPS and Syslog to protect hosting systems from various types of network attacks (internal and external attacks) to lock down servers and allow specific traffic only.
- *Protection from malicious code*: The proposed architecture provides also a supervised machine learning anomaly detection using random forest that protects ASCM from malicious code that could affect pre-established policy and smart contracts. This system permits also to ensure integrity and confidentiality verification during the transmission process.

In practice, the proposed blockchain architecture enables the tracking of all data about products in a supply chain, including product origin and manufacturer, product components and raw materials, and third parties and intermediaries who have intervened in the transportation process. For example, the provenance of a product can be checked using an RFID tag or a QR code, which allows the product to be tracked through every step of the supply chain. In our model, through QR code and GPS technology, blockchain allows the tracking of materials such as aircraft maintenance products and spare parts. This improves visibility and transparency throughout the supply chain.

5 Conclusion

Data security and transparency have recently become key concerns for all businesses, including the airline industries. Thus, blockchain technology has a significant impact on the future of the airline industry and could contribute to increased supply chain traceability and security. Indeed, in the context of airlines, the maintenance process becomes more complex as the number of planes and flights increases. In this context, the most important promise of blockchain is improving data security and ensuring transparency for digital records used throughout the airline supply chain process.

The main purpose of this work is to propose a secure model for records traceability in airline supply chain based on blockchain and machine learning for ASCM context. This framework ensures multi-authentication for all users in the airline supply chain. In addition, our solution permits to manage transaction using blockchain-based consensus process while ensuring accessibility, visibility, transparency, and traceability for all records in ASCM context. Thus, the proposed application permits handling all transactions, checking their validity, and tracking materials recording and GPS data. Furthermore, we propose a security module for anomaly detection based on the use of machine learning. In this regard, the random forest model is used as an ensemble method of the decision tree model. The obtained results show that the significant accuracy rates of the test set are 97%, and the recall rates of the test set are 80%. These results demonstrate the ability of our model to reduce the impact of advanced and persistent attacks.

Like any other research, our research has many limitations that must be acknowledged. First, simulations were performed in a restricted and controlled test environment. Secondly, the framework only allows to manage and secure transaction for the ASCM environment. In future work, we will aim to implement our framework in an uncontrolled environment and test the model in other supply chain contexts.

References

Alshurideh, M. A. (2019). Supply chain integration and customer relationship management in the airline logistics. *Theoretical Economics Letters, 9*, 392.

Appelhanz, S. V.-S. (2016). Traceability system for capturing, processing and providing consumer-relevant information about wood products: System solution and its economic feasibility. *Journal of Cleaner Production, 110*, 132–148.

Azzi, R. R. (2019). The power of a Blockchain-based supply chain. *Computers & Industrial Engineering, 135*, 582–592.

Ball, M. B. (2007). Air transportation: Irregular operations and control. In *Handbooks in operations research and management science*. Elsevier, pp. 1–67.

Belhadi, A. K. (2020). The integrated effect of big data analytics, lean six sigma and green manufacturing on the environmental performance of manufacturing companies: The case of North Africa. *Journal of Cleaner Production, 252*, 119903.

Belhadi, A. M. (2021). Artificial intelligence-driven innovation for enhancing supply chain resilience and performance under the effect of supply chain dynamism: An empirical investigation. *Annals of Operations Research*, 1–26.

Bocek, T. B. (2017). Blockchains everywhere – A use-case of blockchains in the pharma supply-chain. In *IFIP/IEEE symposium on integrated network and service management (IM)*, pp. 772–77.

Bode, C. A. (2015). Structural drivers of upstream supply chain complexity and the frequency of supply chain disruptions. *Journal of Operations Management, 36*, 215–228.

Bodkhe, U. S. (2020). Blockchain for industry 4.0: A comprehensive review. *IEEE Access, 8*, 79764–79800.

Di Vaio, A. (2020). Blockchain technology in supply chain management for sustainable performance: Evidence from the airport industry. *International Journal of Information Management, 52*, 102014.

Earley, K. (2013). Supply chain transparency: forging better relationships with suppliers. *The Guardian*, p. 9.

Ertogral, K. (2019). An integrated production scheduling and workforce capacity planning model for the maintenance and repair operations in airline industry. *Computers & Industrial Engineering, 127*, 832–840.

Flynn, B. P. (2018). Survey research design in supply chain management: The need for evolution in our expectations. *Journal of Supply Chain Management, 54*, 1–15.

Francisco, K. A. (2018). The supply chain has no clothes: Technology adoption of Blockchain for supply chain transparency. *Logistics, 2*, 2.

Kamble, S. S. (2021). A machine learning based approach for predicting blockchain adoption in supply chain. *Technological Forecasting and Social Change, 163*, 120465.

Khan, S. A. (2021). Evaluating barriers and solutions for social sustainability adoption in multi-tier supply chains. *International Journal of Production Research, 59*, 1–20.

Kirejczyk, M. K. (2017). *Ambrosus white paper.* Récupéré sur Ambrosus: https://ambrosus.com/assets/Ambrosus-White-Paper-V8-1.pdf

Kshetri, N. (2017). Blockchain's roles in strengthening cybersecurity and protecting privacy. *Telecommunications Policy, 41*, 1027–1038.

Madhwal, Y. (2017). *Industrial case: Blockchain on aircraft's parts supply chain management.* In American conference on information systems 2017 workshop on smart manufacturing proceedings, pp. 1–6.

Mehra, A. (2020). *Aviation blockchain market by end market- global forecast to 2025.* markets and markets. online https://www.marketsandmarkets.com/pdfdownloadNew.asp?id=175218087

Queiroz, M. M. (2019). Blockchain and supply chain management integration: A systematic review of the literature. *Supply Chain Management: An International Journal, 25*(1), 1–15.

Rejeb, A. K. (2021). Potentials of blockchain technologies for supply chain collaboration: A conceptual framework. *The International Journal of Logistics Management, 176*, 1950–1959.

Santonino III, M. D. (2018). Modernizing the supply chain of Airbus by integrating RFID and blockchain processes. *International Journal of Aviation, Aeronautics, and Aerospace, 4*.

Sawik, B. (2020). Selected multiple criteria supply chain optimization problems. In: *Applications of management science.* Emerald.

Sebbar, A. Z. (2020). MitM detection and defense mechanism CBNA-RF based on machine learning for large-scale SDN context. *Journal of Ambient Intelligence and Humanized Computing, 1*(1), 1–20.

Tasatanattakool, P. (2018). *Blockchain: Challenges and applications.* In 2018 International Conference on Information Networking (ICOIN), pp. 473–475.

Westerkamp, M. F. (2018). *Blockchain-based supply chain traceability: Token recipes model manufacturing processes.* In IEEE International Conference on Internet of Things (IThings) and IEEE Green Computing and Communications (GreenCom) and IEEE Cyber, Physical and Social Computing (CPSCom) and IEEE Smart Data (SmartData), pp. 1595–1602.

Zambre, D. (2013). *Analysis of bitcoin network dataset for fraud.* Unpublished report, 2013.

Zheng, Z. X. (2018). Blockchain challenges and opportunities: A survey. *International Journal of Web and Grid Services, 14*, 352–375.

The Role of IoT and IIoT in Supplier and Customer Continuous Improvement Interface

**Vimal Kumar, Nagendra Kumar Sharma, Ankesh Mittal,
and Pratima Verma**

Abbreviations

AI	Artificial intelligence
B2B	Business to business
B2C	Business to customer
FDI	Foreign direct investment
HR	Human resource
IA	Intelligent asset
IIoT	Internet of Things
IoT	Industrial Internet of Things
PDCA	Plan do check act
QFD	Quality function deployment

V. Kumar (✉)
Department of Information Management, Chaoyang University of Technology, Taichung,
Taiwan
e-mail: vimalkr@gm.cyut.edu.tw

N. K. Sharma
Department of Business Administration, Chaoyang University of Technology, Taichung,
Taiwan

Department of Management Studies, Graphic Era (Deemed to be University), Dehradun, India

A. Mittal
Department of Mechanical Engineering, Asra College of Engineering and Technology, Sangrur,
Punjab, India

P. Verma
Department of Strategic Management, Indian Institute of Management Kozhikode, Kozhikode,
Kerala, India
e-mail: pratima@iimk.ac.in

© The Author(s), under exclusive license to Springer Nature Switzerland AG 2023
S. S. Kamble et al. (eds.), *Digital Transformation and Industry 4.0 for Sustainable
Supply Chain Performance*, EAI/Springer Innovations in Communication and
Computing, https://doi.org/10.1007/978-3-031-19711-6_7

SAP   Systems, applications, and products
SCM   Supply chain management
SME   Small and medium enterprises
TQM   Total quality management
VoC   Voice of the customer
VoP   Voice of producer

1 Introduction

The production process starts from the supplier/producer to make a product or service with different processes. Process improvement is a sequence of actions taken to reach a specific goal. It is imperative to have this improvement process controlled by the PDCA cycle and the voice of customers and voice of producers, that is the need to improve the overall production system in the present competitive era. For enterprises devoted to growth and survival, the TQM (total quality management) strategy is both a practical working procedure and a quality mindset. The TQM strategy begins with the belief that focused management action may improve the organization's service and product quality at a low cost while meeting customer needs. A business needs instantaneous visibility into production, quality, cycle times, machine status, and other critical operational variables to achieve optimum and effective operations (Sishi & Telukdarie, 2020).

The IoT is a network of physical items that have software integrated in them. It's a data-gathering and data-exchanging network of interconnected sensors and electrical gadgets. We all familiar with digital voice-based assistants like Alexa, as well as smart TVs, lightbulbs, and thermostats in our homes? (Peranzo, 2020). In the same way, IIoT, also known as Industry 4.0 (ur Rehman et al., 2019), has machines on the factory floor that are connected via the Internet. The system gathers information in order to power AI and do predictive analytics (Peranzo, 2020). The Industrial Internet of Things (IIoT) is drastically changing the manufacturing business (Peranzo, 2020). It is assisting in the transformation of how items are manufactured and delivered. According to a poll conducted in 2019, around 63% of manufacturers estimate that deploying IoT will boost profitability over the next 5 years. In comparison to 2015, when global IoT spending in discrete manufacturing was 10 billion dollars, it has climbed fourfold to 40 billion dollars in 2020 (Peranzo, 2020). The IIoT aspires to transform industry management and business processes by boosting manufacturing technology products through field data collecting and analysis, resulting in real-time digital twins of industrial settings (Belli et al., 2019). It exhibits the mechanisms for integrating business systems, industrial systems, and processes using IIoT technology (Sishi & Telukdarie, 2020). This chapter demonstrates these IoT and IIoT technologies as the mechanisms for combining business and manufacturing systems and the continuous improvement processes (PDCA cycle).

The PDCA cycle consists of a plan, do, check, and act. The term "plan" refers to a procedure that is defined and documented and has quantifiable goals attached to it. "Do" runs the process and collects the relevant data, while "check" analyzes the data and formats it appropriately. The "act" obtains corrective action and accesses plans using TQM methodologies and processes. Each cycle establishes the standardized pattern based on the analysis.

Voice of producer (VoP) is a term that describes the manufacturing of products and provides various services to the customer, while voice of the customer (VoC) is a phrase that refers to a customer's feedback on their product or service experiences and expectations. It focuses on the requirements, expectations, and understandings of customers, as well as product enhancement (Qualtrix, 2021). These two terms deal with B2B and B2C and managed them. Both B2B and B2C focus on the product's features, services, and all of the interconnections with the IoT environment's parts. IIoT systems emerged during the early stages of a new industrial development that is also based on IIoT service providers and their beneficiaries or consumers (Adner, 2017). In the marketing sphere, significantly greater emphasis is placed on the use of IoT in new techno marketing situations, which improves organizational efficiency in dealing with the B2C market (Tang & Ho, 2019). The customization idea efficiently supports B2B and B2C business systems where digitization and servitization are enabled by IIoT and IoT. It helps producers make swift decisions regarding product enhancement based on timely feedback from customers. Overall, this study is intended to enhance the information flow, real-time connections, decision-making process, customization, production quality improvement programs, supporting data analytics, standardization, globalization, and SCM process improvement with the PDCA cycle in the organization.

This chapter is structured as follows: Sect. 2 provides a detailed background of the present scenario to the development of the background of IoT and IIoT with B2B and B2C, respectively, in supplier and customer interface. This section also highlights the outline of the PDCA cycle with people, plant and machinery, and material methods environment followed by the building the interface model. In Sect. 3 outlines the discussion and conclusions.

2 Scenario and Background with Building the Interface Model

The scenario and background include an in-depth examination of a TQM process with PDCA cycle, VoP and VoC, IIoT and B2B, and IoT and B2C, which are extracted in this section for the current study.

2.1 TQM Process with PDCA Cycle

Quality is a continual process that can be disrupted at any point in the supply and customer service system. The organization can drive their employees and supplies to constantly give quality by letting everyone know how their actions will help satisfy client expectations. They must recognize that they will have internal customers as well as external clients within the organization. In general, a process aids in the transformation of a set of inputs into desired outputs such as products and services. An in-depth examination of the organization's inputs and outputs can aid in determining the best course of action for quality improvement. The "Deming Wheel," on which the goal for continual quality improvement is based, is well-known (Moen & Norman, 2006; Johnson, 2016; Marmara, 2020). The "Deming Wheel" is also known as Shewhart cycle (Moen & Norman, 2006; Johnson, 2016; Marmara, 2020). It follows a PDCA cycle in a certain direction towards achieving the output. Figure 1 shows the PDCA cycle procedure.

The PDCA cycle procedure has been recognized by the following steps: (a) plan: recognize an opportunity and devise a strategy to capitalize on it; (b) do: put the newfound knowledge to the test and investigate the situation on a small scale; (c) check: run the test again, this time examining the results and determining what we have learned; and (d) act: take action based on what we learned in the study step. If the modification isn't successful, we'll have to start over with a different plan. If we passed the exam, we should use what we've learned to make more important changes. Restart the loop by planning new improvements based on what we've learned. It is applied in all manufacturing and service sector organizations where the

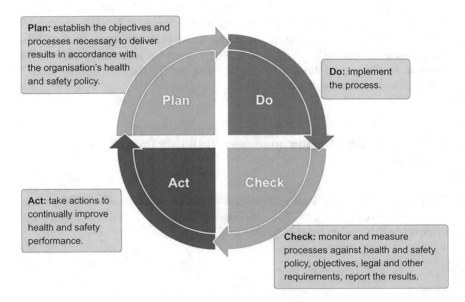

Fig. 1 Plan-do-check-act cycle procedure. (Source: Marmara, 2020)

PDCA cycle provides the roots in scientific methods that lead to performance management (Realyvásquez-Vargas et al., 2018; Du et al., 2008). Further, the performance management aligns with the organization's strategic goal to make their employee very efficient and focus on long-term development which consists of performance plan, implementation, appraisal, and disposal of the outcome (Du et al., 2008). Realyvásquez-Vargas et al. (2018) have applied this PDCA cycle to reduce the number of defects and improve the process continuously. Nowadays, SMEs set out the PDCA cycle in the right direction to improve their productivity and product standards (Jagusiak-Kocik, 2017; Chakraborty, 2016) by continuous strategic learning of the PDCA cycle (Pietrzak & Paliszkiewicz, 2015).

2.2 Voice of Producer and Voice of Customer

The term voice of producer (VoP) figures out what kind of product manufacturing, maintaining the production level and service providers to their customers. It manages the supply of raw materials and overall production and delivery to the market. It deals with the necessary changes after receiving feedback from the customers in the equipment and employee skills to make the process of production level fast and to improve the quality level. At this level, the producers make swift decisions regarding product enhancement based on timely feedback from customers. Only the digitization and servitization enabled by IIoT and IoT make this possible. Furthermore, this customization idea efficiently supports B2B and B2C business systems.

The phrase "voice of the customer" (VoC) refers to the customers' comments on their experiences and expectations for products or services (Crawford, 2008). It focuses on client requirements, expectations, comprehension, and product enhancement. In the words of Akao and King (1990) and Griffin and Hauser (1993), it is an effective approach for discovering unmet requirements and turning them into feasible solutions. According to Bharadwaj et al. (2012), this method is to capture product and process innovation in a supplier-buyer relationship. Numerous methods are associated with VoC ethnography, focus groups, lead-user analysis, customer visit teams, and focus groups to identify customer problems (Cooper & Dreher, 2010). This concept is applicable in business and technology in both study types qualitative and quantitative. Qualitative VoC is often what consumers require and desire, but quantitative VoC demonstrates how customers prioritize their requirements and desires (Aguwa et al., 2012). This method is also applicable in new product development and existing product improvement. They're typically carried out at the outset of any new product, process, or service design endeavor to better understand the customer's wants and needs, as well as the key input for new product definition, quality function deployment (QFD), and the creation of precise design specifications. VoC is also a way to increase customer satisfaction through the measure of the strengths and weaknesses of a business. VoC also helps in strategy development. It accomplishes this by identifying important success elements and key metrics, allowing the organization to be flexible and responsive to client requirements, resulting in a genuine competitive edge (Found & Harrison, 2012). Customer voice behaviors have positive contributions to both customers and service providers (Bove & Robertson, 2005). The suppliers take supply decisions on VoP to improve the process, while producers make decisions based on VoC, the customers' requirement to fulfill their expectations.

2.3 Industrial Internet of Things (IIoT) and B2B

Government policies on Industry 4.0, Smart Factories, Make in India, Make in China 2025, Smart Cities, and Japan's Industrial Value Chain Initiative Forum, as well as enlightened support for green initiatives, rising energy and crude oil prices, favorable FDI, regulatory bodies, and other factors, have propelled the IIoT to its current favorable state (Bortolini et al., 2017; Yang & Gu, 2021). The IIoT cites the expansion and use of IoT in industrial segments and applications (Schneider, 2017; Boyes et al., 2018; Liu et al., 2019; Abuhasel & Khan, 2020). The typical IIoT system is mostly the combination of cloud computing, computerization, machine learning, big data technologies, wireless equipment, cyber-physical systems, internet services, along with others. The aim of the IIoT is also not to completely substitute human-related work but to boost and optimize it (Zhou et al., 2017; Endres et al., 2019; Abuhasel & Khan, 2020). IIoT uses automation and real-time analytics to take benefit of the data that silent machinery has created in industrial settings for years. The driving philosophy behind IIoT is that automatic machines are not only enhanced than humans at confining and evaluating data in real-time, but they're also superior at talking main information that can be used to drive business decisions faster and more accurately (Ambika, 2020; Abuhasel & Khan, 2020).

There is a distinct connection of the emerging platform based on the IIoT and business-to-business (B2B) systems. But, still, there is a lack of research based on the IIoT and B2B and which requires more attention among the researchers (Petrik & Herzwurm, 2019). In the present scenario, the Industrial Internet of Things (IIoT) is emerging in the new industrial setup. It develops a system towards supporting and integrating all the elements of the stakeholders towards adding value (Hein et al., 2019). These platforms embedded with digital capabilities in IIoT enable the systems for customization and value addition services with the aid of external parties (Ardolino et al., 2018). But, these supportive and big platform is missing with giant companies and due to which it is difficult to integrate with the IIoT. This lack of integration hinders the functionality of IIoT and unsupported the other parties to connect with the system (Petrik & Herzwurm, 2019). The IIoT helps in connecting with the company, customers, and the rest of the stakeholders with the help of third-party providers consisting of the system made of hardware and software (Schreieck et al., 2019). IIoT systems evolved in the initial phase of the new industrial development which is also based on the IIoT service providers and its beneficiaries or users (Adner, 2017). The companies (service providers of IIoT) develop an atmosphere for successful functioning by connecting with the other required companies to establish a B2B system. The crux behind the successfulness of IIoT and establishing healthy B2B connections are data based on business relations. It has been noticed and experienced by the experts that the emergence of technologies with digitalization has brought the major transition (Pagani & Pardo, 2017). The IIoT emergence in the business system has modified the business model extensively as it gives more information about the customers, suppliers, and manufacturers (Ahuett-Garza & Kurfess, 2018). A business house that is engaged in manufacturing tyres needs another business house that can buy tyres; this kind of relationship is known as the B2B relationship (Alexander et al., 2009). This relationship is not new but the IIoT can make this connection stronger with the flow of information in the right direction and at the right time (Menon et al., 2019). The relationship of B2B in the market sometimes is more complex specifically while catering to the demand of the business houses from the other one (Sanchez-Iborra & Cano, 2016). Hence, the IIoT integration in this relationship enhances the profitability and improves the services for the other stakeholders in the system (Stock & Seliger, 2016).

Intelligent assets, a data communication infrastructure, analytics and applications to analyze and act on data, and people are the four core components of any IIoT system – (i) intelligent assets (IA): IA management is SAP's next-generation cloud-based collaborative product suite, which enables manufacturers and asset operators to develop, plan, and monitor asset and product service and maintenance strategies through optimal cooperation, integration, and analytical insights (Puri et al., 2020); (ii) communications infrastructure: the technology, products, and network connections that allow for the transmission of communications over long distances are referred to as communications infrastructure, and the category is constantly evolving as new 5G and converged infrastructure capabilities emerge, enabling the use of cutting-edge technologies to power business innovation (Wu et al., 2020); (iii) analytics and applications: automated and remote equipment management and monitoring, predictive maintenance, faster implementation of improvements, pinpoint inventories, quality control, supply chain optimization, and plant safety improvement. With this background, the purpose of this chapter is to go over the IIoT ideas, methods, and protocols in detail; (iv) People: This component is very important among all four components. People are needed to shape the data and determine which metrics have the most effect on the company's bottom line (Boyes et al., 2018). With this background, the purpose of this book chapter is to go over the IIoT ideas, methods, and protocols in further depth.

2.4   Internet of Things (IoT) and B2C

In the present scenario of the business world, the technology-driven market is emerging, and due to the emergence of the internet of things (IoT), it becomes more advance than ever before (Ekren et al., 2021). The IoT was given the power of customization of the products, and it satisfies the customer needs more adequately

(Rymaszewska et al., 2017). The IoT plays a crucial role in maintaining and managing business-to-customer (B2C) setup towards achieving business efficiency. Several studies are presently published, and many are upcoming on IoT and the changing environment of the business (Chan, 2015). The objectives of IoT include but are not limited to the identification, communication, sensing, and collection of data (Oriwoh et al., 2013). Both B2B and B2C focus on the product's features, services, and all the linkages engaged with the elements used within the IoT environment. In the marketing domain, there is much more focus given on engagement of IoT in emerging techno marketing scenarios which enables the organizational efficiency towards handling the B2C market (Tang & Ho, 2019). In terms of the categorization of the IoT market, there are three classes that were explained as B2C (based on the connections of gadgets and people), B2B (based on the connections of associated industries), and business to business to customer (B2B2C) that represents the connections based on the smart cities and smart grids (de Senzi Zancul et al., 2016). The IoT enhances the performance with the integration of digital technologies to even those products which were not connected to the digital systems previously (Turber et al., 2014). The IoT has not only improved the market but also has done the servitization of the manufacturing sections which further gives a boost to the B2C business systems (Lin et al., 2010). These manufacturing sectors are mainly automobile segments, smartphone developments, and all the consumer electronics. As discussed previously that IoT supports enhancing the value of the product through customized work for the consumers. (Takenaka et al., 2016). The use of IoT in the latest consumer products such as smart appliances and consumer durables is making new achievements in product advancements (Han & Lim, 2010). Moreover, there are several other uses of IoT also taking place to make the products closer to the need of consumers (Takenaka et al., 2016). To make certain developments based on the IoT, the manufacturing units are engaged in the data-driven developments of the product (Vancza et al., 2011). The data-based development of the consumer products will enhance the user experience, and it will make the B2C model stronger supported by the user data (Chien et al., 2016). The IoT has a strong upcoming future in the development of the B2C business model in near future with the growth and development of industrialization (Li & Xiao, 2021). Thus, the interface model has been developed considering all the factors and steps of the supplier to customer that reflects the overall model. Figure 2 represents the role of IoT and IIoT in supplier and customer continuous improvement interface.

3   Discussion and Conclusions

TQM is a management method and basic understanding of the principle that involves all employees' dedication to sustaining excellent work standards in all aspects of a company's operations (Marmara, 2020). The TQM process follows the PDCA cycle between producer/supplier and product/service to accomplish the effective output from the established aim and continuous improvement performance (Moen &

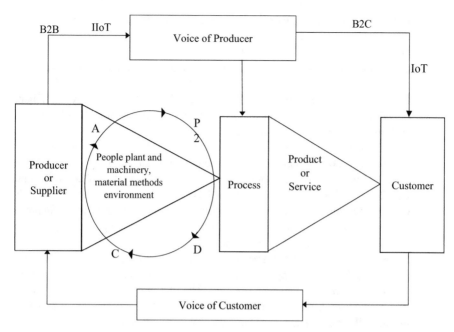

Fig. 2 The IoT and IIoT in supplier and customer continuous improvement interface

Norman, 2006). The involvement of people, plant and machinery, material methods environment plays an important role between producer/supplier and product/service to improve the process continuously. Every time, it needs to check and change the setup with another plan if it does not work and focus on continuous strategic learning of the PDCA cycle (Pietrzak & Paliszkiewicz, 2015). This process leads and is improved by considering oP and VoC, and along with the VoC, the VoP is valuable as well. The result of VoC passes in producer, then producer manufacture products and service according to the results of VoC. The fundamental concept of IIoT refers to the expansion and deployment of the IoT in industrial areas and applications (Schneider, 2017; Boyes et al., 2018; Liu et al., 2019; Abuhasel & Khan, 2020). The output of this study is highlighted in the eight following points:

I. Information flow: It works as a catalyst between the producer and the customer which enhances the flow of information in rapid mode. The introduction of the IIoT into the business system has significantly altered the business model by providing more information on customers, suppliers, and manufacturers (Ahuett-Garza & Kurfess, 2018). This relationship isn't new, but the IIoT may strengthen it by allowing information to flow in the right way and at the appropriate time (Menon et al., 2019). The identification, connectivity, sensing, and gathering of data are the source of information that find a few of the IoT's goals (Oriwoh et al., 2013).

II. Real-time connections: It enables the real-time connection between the factors of production and the external environment which helps to make the production decision prompt. The IIoT theory is that autonomous machines are not only better than people at storing and interpreting data in real-time, but they're also better at communicating key information that can be used to make faster and more accurate business decisions (Ambika, 2020; Abuhasel & Khan, 2020). Data based on business relationships is at the heart of IIoT's success and the establishment of healthy B2B partnerships. The network connections that allow for the transmission of communications over long distances are referred to as communications infrastructure (Wu et al., 2020). There are three types of connections between business and customer such as B2C, B2B, and B2B2C, which represent connections based on smart cities and smart grids (de Senzi Zancul et al., 2016). For instance, the connection between Siemens and MindSphere integrates their partnership and developed ecosystem in distinctive phases (Petrik & Herzwurm, 2019).

III. Decision-making process: The implementation of digitalization backed by IIoT and IoT gives the opportunity to the production manager towards making various decisions in the production unit. The PDCA cycle provides the way to make a decision but improves the arrangement to take a better decision where the overall model makes faster and more accurate business decisions.

IV. Production quality improvement programs: Based on prompt feedback from a customer, the producers take quick decisions regarding the improvement of the product. It is possible only on the digitization and servitization backed by IIoT and IoT. At both levels, the PDCA cycle offers to the management system to follow step by step process to improve the product quality.

V. Supporting data analytics: The concept of IIoT and IoT enables the entire business system externally and internally towards generating and converting data into meaningful information for decision-makers. Big data technology is used in IIoT systems to take advantage of the data that silent machinery has generated in industrial settings for years through automation and real-time analytics. Manufacturing units are involved in data-driven product creation to create certain developments based on IoT (Vancza et al., 2011).

VI. Customization: The biggest role of the present technology based on IIoT and IoT helps in catering to the changing demand of the consumers and customers more adequately than ever before. Further, this concept of customization supports B2B and B2C business systems effectively. These IIoT platforms with digital capabilities enable systems to be customized, and value-added services can be provided by third parties (Ardolino et al., 2018), and the IoT has been given the ability to customize items, allowing it to better meet the wants of customers (Rymaszewska et al., 2017).

VII. Standardization and globalization: The real-time data and information help in product standardization as per the international standard. It also helps in presenting the firm or company at the international level.

VIII. SCM process improvement with PDCA cycle: The supply chain between supplier and customer is managed by step by step process. Managing the process is not so easy to shift the employees to equipment. As a result of improved productivity, the best method to ensure that the equipment runs smoothly is to work with trained employees. The supply chain management process is influenced and improved by the combined effect of the PDCA cycle and the process.

The overall TQM provides the three support systems such as technological support, human support, and cultural support. Technological support is concerned with the process of designing and building quality into a product/service. Human support is concerned with empowering people to demonstrate mastery over the task at hand and to take action that produces useful consequences. It makes it easier to employ teamwork to improve the necessary changes. Cultural support is concerned with the organizational environment that fosters quality-mindedness. The combined effect of these three support systems manages quality to boost and the organization environment where the PDCA cycle deals with the continuous improvement. As a result, technological, human, and cultural support covers the majority of the process and eliminates all financial, human resource, and skill set obstacles. The three support systems of TQM practices are shown in Fig. 3.

In recent years, a number of rising technologies like big data, cloud computing, machine learning, artificial intelligence, and wireless networks convey superior chances for encouraging industrial upgrades and allowing the fourth industrial revolution (Piccialli et al., 2021). These technologies enhance operational efficiency, productivity, and visibility and also reduce the complications of the operational process. The IIoT supports the fourth industrial revolution with transformative manufacturing and production strategies by connecting the shop floors with a cloud platform to get the status of raw material progress in real-time (Cakir et al., 2021).

Fig. 3 The three support systems of TQM practices

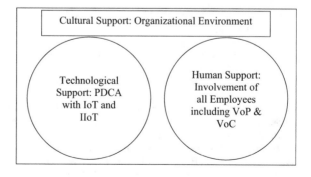

References

Abuhasel, K. A., & Khan, M. A. (2020). A secure industrial internet of things (IIoT) framework for resource management in smart manufacturing. *IEEE Access, 8*, 117354–117364.

Adner, R. (2017). Ecosystem as structure: An actionable construct for strategy. *Journal of Management, 43*(1), 39–58.

Aguwa, C. C., Monplaisir, L., & Turgut, O. (2012). Voice of the customer: Customer satisfaction ratio based analysis. *Expert Systems with Applications, 39*(11), 10112–10119.

Ahuett-Garza, H., & Kurfess, T. (2018). A brief discussion on the trends of habilitating technologies for industry 4.0 and smart manufacturing. *Manufacturing Letters, 15*, 60–63.

Akao, Y., & King, B. (1990). *Quality function deployment: Integrating customer requirements into product design* (Vol. 21). Productivity press.

Alexander, N. S., Bick, G., Abratt, R., & Bendixen, M. (2009). Impact of branding and product augmentation on decision making in the B2B market. *South African Journal of Business Management, 40*(1), 1–120.

Ambika, P. (2020). Machine learning and deep learning algorithms on the Industrial Internet of Things (IIoT). *Advances in Computers, 117*(1), 321–338.

Ardolino, M., Rapaccini, M., Saccani, N., Gaiardelli, P., Crespi, G., & Ruggeri, C. (2018). The role of digital technologies for the service transformation of industrial companies. *International Journal of Production Research, 56*(6), 2116–2132.

Belli, L., Davoli, L., Medioli, A., Marchini, P. L., & Ferrari, G. (2019). Toward industry 4.0 with IoT: Optimizing business processes in an evolving manufacturing factory. *Frontiers in ICT, 6*, 17.

Bharadwaj, N., Nevin, J. R., & Wallman, J. P. (2012). Explicating hearing the voice of the customer as a manifestation of customer focus and assessing its consequences. *Journal of Product Innovation Management, 29*(6), 1012–1030.

Bortolini, M., Ferrari, E., Gamberi, M., Pilati, F., & Faccio, M. (2017). Assembly system design in the industry 4.0 era: A general framework. *IFAC-PapersOnLine, 50*(1), 5700–5705.

Bove, L. L., & Robertson, N. L. (2005). Exploring the role of relationship variables in predicting customer voice to a service worker. *Journal of Retailing and Consumer Services, 12*(2), 83–97.

Boyes, H., Hallaq, B., Cunningham, J., & Watson, T. (2018). The industrial internet of things (IIoT): An analysis framework. *Computers in Industry, 101*, 1–12.

Cakir, M., Guvenc, M. A., & Mistikoglu, S. (2021). The experimental application of popular machine learning algorithms on predictive maintenance and the design of IIoT based condition monitoring system. *Computers & Industrial Engineering, 151*, 106948.

Chakraborty, A. (2016). Importance of PDCA cycle for SMEs. *SSRG International Journal of Mechanical Engineering, 3*(5), 30–34.

Chan, H. C. (2015). Internet of things business models. *Journal of Service Science and Management, 8*(04), 552.

Chien, C. F., Kerh, R., Lin, K. Y., & Yu, A. P. I. (2016). Data-driven innovation to capture user-experience product design: An empirical study for notebook visual aesthetics design. *Computers & Industrial Engineering, 99*, 162–173.

Cooper, R. G., & Dreher, A. (2010). Voice-of-customer methods. *Marketing management, 19*(4), 38–43.

Crawford, C. M. (2008). *New products management*. Tata McGraw-Hill Education.

de Senzi Zancul, E., Takey, S. M., Barquet, A. P. B., Kuwabara, L. H., Miguel, P. A. C., & Rozenfeld, H. (2016). Business process support for IoT based product-service systems (PSS). *Business Process Management Journal, 22*(2), 305–323.

Du, Q. L., Cao, S. M., Ba, L. L., & Cheng, J. M. (2008, October). *Application of PDCA cycle in the performance management system*. In 2008 4th International conference on wireless communications, networking and mobile computing (pp. 1–4). IEEE.

Ekren, B. Y., Mangla, S. K., Turhanlar, E. E., Kazancoglu, Y., & Li, G. (2021). Lateral inventory share-based models for IoT-enabled E-commerce sustainable food supply networks. *Computers & Operations Research, 130*, 105237.

Endres, H., Indulska, M., Ghosh, A., Baiyere, A., & Broser, S., (2019). *Industrial internet of things (IIoT) business model classification.* In 40th International conference on information systems, ICIS 2019 (p. 2988). Association for Information Systems. AIS Electronic Library (AISeL).

Found, P., & Harrison, R. (2012). Understanding the lean voice of the customer. *International Journal of Lean Six Sigma, 3*(3), 251–267.

Griffin, A., & Hauser, J. R. (1993). The voice of the customer. *Marketing Science, 12*(1), 1–27.

Han, D. M., & Lim, J. H. (2010). Design and implementation of smart home energy management systems based on zigbee. *IEEE Transactions on Consumer Electronics, 56*(3), 1417–1425.

Hein, A., Weking, J., Schreieck, M., Wiesche, M., Böhm, M., & Krcmar, H. (2019). Value co-creation practices in business-to-business platform ecosystems. *Electronic Markets, 29*(3), 503–518.

Jagusiak-Kocik, M. (2017). PDCA cycle as a part of continuous improvement in the production company-A case study. *Production Engineering Archives, 14*, 19–22.

Johnson, C. N. (2016). The benefits of PDCA. *Quality Progress, 49*(1), 45.

Li, Q., & Xiao, R. (2021). The use of data mining technology in agricultural e-commerce under the background of 6G internet of things communication. *International Journal of System Assurance Engineering and Management*, 1–11.

Lin, Y., Shi, Y., & Zhou, L. (2010). *Service supply chain: Nature, evolution, and operational implications.* In Proceedings of the 6th CIRP-sponsored international conference on digital enterprise technology (pp. 1189–1204). Springer.

Liu, M., Yu, F. R., Teng, Y., Leung, V. C., & Song, M. (2019). Performance optimization for blockchain-enabled industrial Internet of Things (IIoT) systems: A deep reinforcement learning approach. *IEEE Transactions on Industrial Informatics, 15*(6), 3559–3570.

Marmara, Z. (2020, September). *Plan-do-check-act and the grandfather of total quality management.* Retrieved August 26, 2021, from https://zmarmara.medium.com/plan-do-check-act-and-the-grandfather-of-total-quality-management-391387e0ac85

Menon, K., Kärkkäinen, H., Mittal, S., & Wuest, T. (2019, July). *Impact of IIoT based technologies on characteristic features and related options of nonownership business models.* In IFIP International conference on product lifecycle management (pp. 302–312). Springer.

Moen, R., & Norman, C. (2006). *Evolution of the PDCA cycle* (pp. 1–11). Retrieved August 26, 2021, from http://citeseerx.ist.psu.edu/viewdoc/download?doi=10.1.1.470.5465&rep=rep1&type=pdf

Oriwoh, E., Sant, P., & Epiphaniou, G. (2013). Guidelines for internet of things deployment approaches–The thing commandments. *Procedia Computer Science, 21*, 122–131.

Pagani, M., & Pardo, C. (2017). The impact of digital technology on relationships in a business network. *Industrial Marketing Management, 67*, 185–192.

Peranzo, P. (2020, September 4). *Digital transformation in manufacturing: 7 trends to watch in 2020.* Retrieved July 29, 2021, from https://www.imaginovation.net/blog/digital-transformation-manufacturing-trends-2020/

Petrik, D., & Herzwurm, G. (2019, November). *Towards an understanding of iIoT ecosystem evolution-MindSphere case study.* In International conference on software business (pp. 46–54). Springer.

Piccialli, F., Bessis, N., & Cambria, E. (2021). Industrial Internet of Things (IIoT): Where we are and what's next. *IEEE Transactions on Industrial Informatics, 99*, 1–1.

Pietrzak, M., & Paliszkiewicz, J. (2015). Framework of strategic learning: The PDCA cycle. *Management (18544223), 10*(2), 149–161.

Puri, V., Priyadarshini, I., Kumar, R., & Kim, L. C. (2020, March). *Blockchain meets IIoT: An architecture for privacy preservation and security in IIoT.* In 2020 International Conference on Computer Science, Engineering and Applications (ICCSEA) (pp. 1–7). IEEE.

Qualtrix. (2021). *What is voice of the customer (VoC)?* Accessed July 29, 2021, from https://www. qualtrics.com/au/experience-management/customer/what-is-voice-of-customer/?rid=ip& prevsite=en&newsite=au&geo=TW&geomatch=au

Realyvásquez-Vargas, A., Arredondo-Soto, K. C., Carrillo-Gutiérrez, T., & Ravelo, G. (2018). Applying the Plan-Do-Check-Act (PDCA) cycle to reduce the defects in the manufacturing industry. A case study. *Applied Sciences, 8*(11), 1–17. 2181.

Rymaszewska, A., Helo, P., & Gunasekaran, A. (2017). IoT powered servitization of manufacturing–An exploratory case study. *International Journal of Production Economics, 192*, 92–105.

Sanchez-Iborra, R., & Cano, M. D. (2016). State of the art in LP-WAN solutions for industrial IoT services. *Sensors, 16*(5), 708.

Schneider, S. (2017). The industrial internet of things (iiot) applications and taxonomy. In *Internet of things and data analytics handbook* (pp. 41–81).

Schreieck, M., Wiesche, M., Kude, T., & Krcmar, H. (2019). *Shifting to the cloud–how SAP's partners cope with the change.*

Sishi, M., & Telukdarie, A. (2020). Implementation of industry 4.0 technologies in the mining industry-A case study. *International Journal of Mining and Mineral Engineering, 11*(1), 1–22.

Soltani, S. (2021). B2B engagement within an internet of things ecosystem. *Journal of Business & Industrial Marketing.*

Stock, T., & Seliger, G. (2016). Opportunities of sustainable manufacturing in industry 4.0. *Procedia Cirp, 40*, 536–541.

Takenaka, T., Yamamoto, Y., Fukuda, K., Kimura, A., & Ueda, K. (2016). Enhancing products and services using smart appliance networks. *CIRP Annals, 65*(1), 397–400.

Tang, T., & Ho, A. T. K. (2019). A path-dependence perspective on the adoption of internet of things: Evidence from early adopters of smart and connected sensors in the United States. *Government Information Quarterly, 36*(2), 321–332.

Turber, S., Vom Brocke, J., Gassmann, O., & Fleisch, E. (2014, May). *Designing business models in the era of internet of things.* In International conference on design science research in information systems (pp. 17–31). Springer.

ur Rehman, M. H., Yaqoob, I., Salah, K., Imran, M., Jayaraman, P. P., & Perera, C. (2019). The role of big data analytics in industrial internet of things. *Future Generation Computer Systems, 99*, 247–259.

Váncza, J., Monostori, L., Lutters, D., Kumara, S. R., Tseng, M., Valckenaers, P., & Van Brussel, H. (2011). Cooperative and responsive manufacturing enterprises. *CIRP Annals, 60*(2), 797–820.

Wu, Y., Dai, H. N., & Wang, H. (2020). Convergence of blockchain and edge computing for secure and scalable IIoT critical infrastructures in industry 4.0. *IEEE Internet of Things Journal, 8*(4), 2300–2317.

Yang, F., & Gu, S. (2021). Industry 4.0, a revolution that requires technology and national strategies. *Complex & Intelligent Systems, 7*(3), 1311–1325.

Zhou, C., Damiano, N., Whisner, B., & Reyes, M. (2017). Industrial internet of things:(IIoT) applications in underground coal mines. *Mining Engineering, 69*(12), 50.

Customer Relationship Management in the Digital Era of Artificial Intelligence

Sheshadri Chatterjee and Ranjan Chaudhuri

1 Introduction

Artificial intelligence (AI) is considered as a concept by which it is possible for a machine to think, share, and even imitate intelligent behavior of humans (Awasthi & Sangle, 2013). Machines embedded with AI can have the processing ability to perform some of the important tasks which include planning, learning, and even realizing language without interference of humans (Baabdullah et al., 2021). The brain that could make AI a prolific source of doing different intelligent works is machine learning (ML). ML is associated with a structured set of algorithms capable of processing data, efficiently learning from data, and utilizing data for accurate decision-making (Basile et al., 2021). The ML process is based on AI technology, and it helps to accelerate human jobs and responsibilities. It helps decision-making faster, accurate, and easier (Maxwell et al., 2011). Thus, the above discussions advocate that AI-embedded technology possesses prolific processing abilities and automates different tasks. In this scenario, AI has been considered as an important and helpful technology to ameliorate customer relationship management (CRM) process, features, and functionalities (San-Martína et al., 2016).

CRM is considered as an effective tool for understanding the customers more accurately as well as intimately by "identifying a company's best customers and maximizing the values from them by satisfying and retaining them" (Kennedy, 2006, p.58). Relationship management is associated with putting more efforts to develop a

S. Chatterjee (✉)
Department of Computer Science & Engineering, Indian Institute of Technology Kharagpur, Kharagpur, West Bengal, India

R. Chaudhuri
Department of Marketing, Indian Institute of Management Ranchi, Ranchi, Jharkhand, India
e-mail: ranjan.chaudhuri@iimranchi.ac.in

direct and close relationship with the potential customers in the business context (Harmeling et al., 2015). Organizational authority could realize the need of knowing more about the customers and their needs for taking congenial initiatives to create values for the customers (Bishop, 2009). During 1980s, this initiative was found lamentably deteriorating as the marketers emphasized more focus on developing products and segmented marketing than to be engaged for collecting data of customers to improve relationship management (Peelen, 2005). Eventually with the introduction of information and communication technology (ICT), the CRM procedure received recognition when marketers achieved good results within a short span of time by successfully processing the customers' data with the help of ICT. In this favorable situation, marketers realized the need of understanding the importance of retaining potential customers to ensure more business profits (Wang et al., 2016; Watson et al., 2015). However, the present era of digitalization has brought change in the organizational business processes and practices enabling the organizations to adopt new production processes, and much of the phenomenon are found to depend on Industry 4.0 applications including Internet of Things (IoT), AI, blockchain technology, and so on (Paulus-Rohmer et al., 2016). The organizations are now profoundly engaged in managing huge volume of data of customers for the betterment of their CRM system (Bhattacharya & Chatterjee, 2020; Tsai et al., 2021). But it is difficult to analyze such huge volume of data of customers accurately in a cost-effective manager within a short time by manual efforts (Li & Nguyen, 2016). The analysis of such huge volume of customers' data by humans is very difficult, and as such, applications of modern technology are needed to ease out the human manual efforts (Molinillo & Japutra, 2017). With the applications of AI, such huge volume of customers' data can be analyzed in an accurate, cost-effective, and quicker way without human intervention (Bose, 2002; Schultz & Pick, 2012).

2 Literature Review

For retaining the potential customers, CRM is considered as a strategic initiative which would ensure profitability to the organizations (Cannon & Perreault, 1999; Heide et al., 2007). To improve CRM process, huge volume of customers' data are needed to be gathered for knowing their daily needs, preferences, purchasing habits, likings, disliking, and so on (Padmanabhan & Tuzhilin, 1999; Raghunathan, 1999). Basing on this information, customers are categorized in different segments or groups (Graca et al., 2015). In this way, CRM leads an organization to be more customer-centric with the help of modern technology like AI (Sahu & Gupta, 2007; San-Martína et al., 2016; Verma & Verma, 2013). A study has shown that from the beginning of 2017 to the end of 2021, the AI-CRM activities will ensure global business revenue to the tune of $1.1 trillion (Gantz et al., 2017). CRM can be defined as "integration of technologies and human processes used to satisfy the needs of a customer during any given interaction. More specifically, CRM involves acquisition, analysis, and usage of knowledge about customers to sell more goods or services and to do it, more efficiently" (Bose, 2002, p.1). Acquisition of multifarious data of customers helps the organizations towards accurate prediction through data analysis. Since the data of customers is huge, many organizations use predictive modelling techniques (Wierenga, 2009). Many organizations now are used to take help of modern technology to manage huge volume of data of customers which are of various nature. Few of the modern techniques used by CRM technology include classification and regression trees (CART) and neural network (NN). These techniques are dependent on AI. For such reasons, the organizations are trying to embed AI with CRM to maximize the profit (Chatterjee, 2019c; Chaudhuri et al., 2021). Organizations are continuously trying to accurately predict the customers' behavior taking help of predictive techniques with support of complex mechanisms using AI technology (Hopkinson et al., 2018). In this way, with the help of AI-integrated CRM system, organizations try to store huge volume of customers' data in the cloud platform. For analysis of such data quickly, accurately, and with incurring less cost, use of AI technology is essential (Wen & Chen, 2010). It is observed that AI has been able to offer new practices and challenges in the B2B marketing prospect (Davenport et al., 2019; Rust, 2020). Studies have also showed that to flourish, the marketers need smart solution for automating the structure of aligning, standardizing, as well as customizing data in the complex business environment, and for this, the role of AI-CRM seemed to be crucial (Akter et al., 2019; Jabbar et al., 2019; Libai et al., 2020). The increasing availability of data of customers integrated with the contribution of AI has helped the organizations to create personalized services which could fetch considerable gain to the customers and organizations (Gupta et al., 2020; Rust & Huang, 2014). Study demonstrated that AI-CRM could help the organizations for improving profit maximization portfolio by improving customer relationship activities (Saura et al., 2019). It is a fact that AI-integrated CRM system is deriving immense benefit to the organizations, but it is known that many organizations are facing several challenges in using AI with their existing legacy CRM system which is standing on the way to ensure performance gain (Fountaine et al., 2019; Hu et al., 2021; Nishant et al., 2020). This is owing to restructuring and implementational issues (Mikalef & Gupta, 2021). Hence, the organizations are needed to develop the complementary resources for the best usage of AI technology and related investments (Bag et al., 2021; Balakrishnan & Dwivedi, 2021; Grover et al., 2020).

3   Digital Transformation of Organizations and Sustainability

Digital transformation is considered as an adoption of digital technology by an organization for improving values of customers, business processes, as well as innovation (O'Donnell, 2017). Digital transformation is interpreted as a process to convert analog information into the digital form. This can be done by the help of analog to digital converter like image scanner as well as sampling of voice (Sakka, Chatterjee, Chaudhuri, & Thrassou, 2021a). Digitalization is construed as an organizational process to ensure technologically induced alteration (Gobble 2018). Digitization in an organization has revolutionized the business process by inducing new business models like electronic payment, business digitalization, and office automation along with paperless office practices (Chatterjee, 2021a; Vrontis et al., 2020). This is being done by using new technologies like big data applications, smartphones, blockchain, business intelligence, cloud services, crypto currencies, and so on (Sakka, Chatterjee, Chaudhuri, & Apoorva, 2021b). Digital transformation has brought great change in the organization, but it has also brought changes in the economy and in the society (Bounfour, 2016; Vrontis, Chatterjee, & Chaudhuri, 2021a). Digitization can be conceptualized in terms of changing of signal processes into digital form which is transmitted with the help of networks or digital services using in-built systems (Vogelsang, 2010; Vrontis et al., 2021b). Digital transformation is nothing but a complete social effect of digitalization (Khan, 2017). Digital transformation in an organization is associated with different levels of intensity. These include different functions in the sales channels concerning with e-commerce, more information and presentation via websites, and business process integration transformed to new business models associated with virtual products and services (Härting et al., 2017). Digitalization in an organization has been ensured by digitization that could help to change the business models, socio-economic structures, patterns of consumption, organizational pattern, and even it influences the cultural barriers (Cochoy & Peterson McIntyre, 2017). Digitalization has transformed the society into digital society (Bowen & Giannini, 2014). There should not be any confusion between digitization and digitalization along with the digital transformation. Technical conversion is associated with the concept of digitization. Consequent upon digitization in organization, the business process undergoes changes which is called digitalization. Its effects are associated with the concept of digital transformation which illuminates and facilitates the already existing processes and practices (Khan, 2017). There was a common wrong understanding that digitalization means more and more usage of IT in organization for enjoying the advantage of digital technology. Now due to ubiquitous effects of digitalization, the concept has assumed wider amplitude because it is now linked with the holistic views on the business with social change. Figure 1 shows different AI-CRM applications and digital transformation in the organizations.

It is observed that most of the organizations are found to emphasize more on technology in the context of digitalization forgetting about the customers (Siachou et al., 2021; Vrontis, Chatterjee, Chaudhuri, Thrassou, Ghosh, & Chaudhuri, 2021c). A successful organization is needed to harness the culture as well as strategy since the goal of digital strategy of an organization should be to improve efficiency, user experience, brush up innovation through dependence on human capital, and increase of accurate decision-making (Chatterjee, 2021b). However, digital transformation in an organization is concerned with gaining lucrative opportunities as well as challenges (Aleksej et al., 2018). The digital transformation brings a cultural change which is opposed by the workers usually and leaders towards adoption and reliance

Fig. 1 AI-CRM applications and digital transformation

with unfamiliar technology (Jane, 2015), but this transformation can create unique marketplace (Reinartz et al., 2018). However, all the organizations are needed to adhere to digital transformation strategy overcoming all the challenges to achieve success towards its sustainability.

4   Integration of AI with CRM

Organization using Industry 4.0 technology like AI, blockchain, and so on has helped in the automation process. Automation has also facilitated the interaction between the potential customers and the brands (Ostrom et al., 2015). Previously CRM technology could ensure such interaction between customers and brands through automatically generated notifications associated with auto-responding (Sreenivasulu & Chatterjee, 2019). But the interference of AI has made these mechanisms easier and more accurate. It is a fact that by the grace of AI, the relationship between potential customers and the organizations has been increased through automated processes. The AI-integrated CRM has simplified the business process in the context of customer relationship management as the usage of AI has been able to do things which were previously done manually (Kumar et al., 2016). CRM helps to improve the retention of the customers and increases the numbers of customers (Grönroos, 1990). Scholars subsequently thought of adding another phase with the abovementioned phases which is termination phase (Grönroos, 1994). However, integration of AI with CRM has been able to accelerate the process of interaction in a smoother and cost-effective manner (Sreenivasulu & Chatterjee, 2021). To attract the customers, AI could help with the customers in a pre-sale period by the features such as chat bots along with intelligent contact marketing, and it can ensure automated responding through emails for supplying multifarious information

Fig. 2 Integration of AI
with CRM

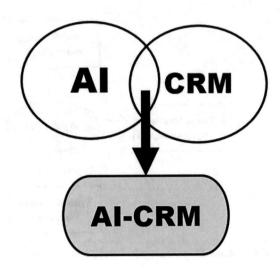

Fig. 2 Integration of AI with CRM

to the potential customers (Chatterjee, 2019d; Nguyen et al., 2019). In the post-sale period, to retain the customers, the AI technology can help to answer the post-sale queries of the customers accurately keeping records of the previous customers accurately. Through predictive analysis, AI can advise the organizations by foreseeing which customers would stay and which customers will go away to the competitors. By the help of intelligent marketing process, AI can extend and deepen the CRM activities to enhance the number of customers (Chatterjee, 2019b). By alerting customers about the information for renewal of contract through automated email services, AI can help customers in their termination phase along with other advantages. In this way, in all the four stages (attraction, retention, enhancement, and termination of customers), AI can improve the CRM activities. The integration of CRM with AI technology is shown in Fig. 2.

5 Application of AI in CRM

AI can be used to improve CRM activities in many ways. AI can help an organization to develop its information management system using predictive modelling technique (Bradlow et al., 2017). AI can help organizational CRM activities in several ways. A few of them are as follows:

- Huge volume of data can be gathered by AI in an efficient and cost-effective way.
- AI can help to process these collected data in an easier way.
- AI can help organizations to improve the business priestesses with the help of natural language processing (NLP) system when necessary.
- AI can help the organizations to realize accurately the behavioral attitude of the customers.

- AI can caution the organizations about the risks that might arise in the process of business growth.
- AI can help the organizations to solve any problem that impedes the growth and can help the organizations for overcoming any untoward risk.

Many studies are available where it has been observed that AI can improve the CRM activities (Syam & Sharma, 2018). Social CRM activities help the organizations to augment CRM activities in an advanced way (Bradlow et al., 2017). Ubiquitous CRM (u-CRM) activities help an organization to provide its customers real-time servicing (Parise et al., 2016). With the help of mobile CRM (m-CRM) activities, an organization can help its potential customers in multiple ways (Culotta et al., 2015). AI technology can help an organization by various ways through improvement of its CRM activities. CRM activities are mainly related with development of relationship between customer and organization. It is a social phenomenon. In the context of professional services, social interactions help to assess the service quality of an organization in the perspective of its CRM activities (Hogg & Gabbott, 1998). However, though it has been observed that AI helps an organization in various ways, scholars believe that there are other things which can be helped by AI in the future. In this context, scholars have opined that "Although AI is already in use by the thousands of companies around the world, most big opportunities have yet not been tapped" (Brynjolfsson & Mcafee, 2017). In this context, it is expected that there exist more opportunities for the organizations to extract more benefits from AI technology in the perspectives of improvement of business relationship in future.

6   AI Algorithm and Related Data Mining

For the appropriate applications of AI-CRM technology, the role of data as well as AI algorithm plays a critical role towards improvement of business activities of the organizations. In the context of ensuring effective and successful AI-CRM applications in organizations, it is necessary to meaningfully create the application through appropriate use of AI algorithm. This includes comparatively small data model, ensuring uninterrupted access towards effective data as well as efficient cloud computing system which should be inexpensive and powerful (Alshare & Lane, 2011). For improving CRM activities, most of the organizations have used AI algorithms along with their products. For examples, Dynamics 365 for customer insights can help organizations as well as customers in many ways such as automated recommendations, opportunity risk scoring, understanding customers' likings and disliking, and so on. AI algorithms can help in predictive analysis with the inputs from various kinds of customers' data. The AI algorithms could be able to bring better results to the organizations (Awasthi & Sangle, 2013). Thus, the organizations should collect latest customers' data, then clean, and curate the data as input to the AI-CRM system and after that based on such appropriate input data, AI algorithm would be used to extract effective and meaningful results which will help business activities of the organizations (Alshibly, 2015).

7   Optimization of Business Operations Using AI-CRM

In this section some specific applications of AI in CRM have been discussed.

7.1   Customer Service and Retention of Customers

Huge amount of data of customers are gathered by the organizations. With the help of AI-integrated CRM system, it is easy and accurate to extract congenial and appropriate insights from such huge amount of data of customers (Chatterjee, 2020). For preparing the target profile of customers, such insights are deemed to be helpful. Once it is possible to identify the target customers by the help of AI-CRM, it is easy for the sales personnel of the organizations to interact with such customers (Dwivedi et al., 2021). This process will help to retain the customers and will also help to increase the satisfaction of the customers. The AI-CRM technology will guide the salespersons regarding the mode of interactions with the target customers – whether through mobile or through email (Thrassou et al., 2021a).

7.2   Automate the Routine Tasks

AI-CRM technology will help the organizations to list out the mandatory routine tasks of the organizations with priority and reducing time. Organizational different activities will be successfully handled by AI-CRM activities in an automated way (Chatterjee, 2019a). These include data input and retrieval, updating forecasts, call list determination, and so on. For accurate recommendations and decision-making, AI-CRM helps the organizations in various ways (Ghosh et al., 2020).

7.3   Guidance to the Sales Team

For accurately planning the roadmap of the organizations, AI-CRM plays a critical role. This will help the organizations towards moving ahead with appropriate steps from lead to deal (Damaševičius et al., 2020). In any type of situation, for achieving better results, the AI-CRM helps a lot to the sales team (Ghosh et al., 2021).

Customer Segmentation	Churn Prediction	Segment Analytics
Service Marketing	Market Basket Analysis	Fraud Detection

Fig. 3 AI-CRM applications in organizations

7.4 Virtual Assistance

AI-CRM may act as a helpful virtual assistant of the organizations. This hybrid system will help for automating customers' response and will help to capture data needed for dealing with the web behavior of the potential customers as well as with their demographics (Kar & Chatterjee, 2020). This system will help to automate emails to the customers and can automatically book appointments with the targeted customers (Ghosh et al., 2019). Figure 3 shows different AI-CRM applications in the organizations.

7.5 Prioritization, Lead Customization, and Appropriate Representation

AI-CRM will help the organizations to understand the habits, likings, disliking, and purchase history of the potential customers (Nguyen et al., 2019). These inputs will help the organizations to appropriately segment the customers into different appropriate categories. This system will help for lead generation and prioritizing (Nguyen & Chatterjee, 2021).

The above discussions help to understand that AI-CRM applications assist a lot to the sales team of an organizations. The AI-CRM helps the team to appropriately build the relationship with the existing and potential customers of the organizations.

8 Examples of AI-CRM

Several organizations have used different software for successfully applying AI-integrated CRM system in their platforms. Sugar CRM introduced AI feature known as "Hint." Zoho introduced an AI functionality named as "Zia." Microsoft has introduced "Dynamics 365," a software with AI features. Salesforce has introduced a software feature named as "Einstein" which is AI enabled.

Sugar CRM This organization has developed a software "Hint." This software can automatically tune, search, as well as provide inputs which are helpful for collecting personal and corporate profile with details (Nguyen et al., 2019; Piccolo et al., 2021). The results come out almost instantaneously.

Zoho This organization has provided a software named as "Zia." It functions as a conversational AI assistant. It can work like simple data analytics tool as well as complex data analysis tool simultaneously. It has been developed in such a way that one can simply talk with this software by calling it as "Zia" using mobile (Nguyen et al., 2019).

Microsoft Microsoft has devised a software name as "Dynamics 365." This software is capable of being involved in effective and accurate predictive analysis with the help of customer data. It has been built to ensure effective relationship taking help of the collected data for providing actionable matrices.

Salesforce This organization has developed a software named as "Einstein." It is helpful for providing prediction with accurate recommendations taking help of data already captured (Rana et al., 2020).

Figure 4 shows different underlying techniques for AI-CRM applications.

Fig. 4 Underlying techniques for AI-CRM applications

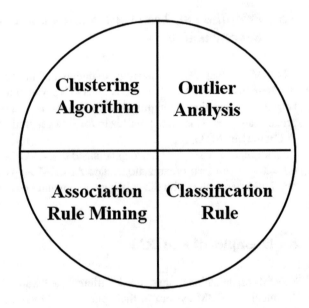

9 Implication of This Study

It has been shown that CRM embedded with AI is a powerful tool to help an organization to improve relationship with the customers. To extract better results with application of AI, correct data are needed to be supplied to this hybrid system to fetch better results. Thus, to deliver better results, AI is found to have depended on correct and curated data. Obviously, in this context, Internet plays a crucial role (Nguyen et al., 2020). To improve modern marketing technique and sales efforts, Internet and social media usage are deemed to be important. The AI-CRM tools always engage to collect customers' information of various types. Now, an employee of an organization not only likes to collect data of customers such as the customers' contact details, but the employee also likes to collect detail statistics of every interaction with the customers and the organization. The employee likes to collect the job hierarchy of the customers, and even they may like to collect data which cover usage history of social media portfolio of the customers (Rana et al., 2021a). However, it is not an easy task by the salespeople to collect such huge volume of customers' data of multifarious nature and to analyze those data. In this context, the organization seeks help of AI-CRM (Thrassou et al., 2021b). This technology (AI-CRM) will help the organizations in many ways. Some salient important features are as follows:

- *Flawless forecasting.*
- *Appropriate recommendations which enhance sales.*
- *Searching as well as social profiling of customers with support of NLP.*
- *Predictive lead scoring which helps for accurate decision-making.*

By getting curated and actionable data of the customers, it is possible to make smart recommendations by the help of AI-CRM. This would derive benefits to the organizations. Thus, by capturing usable data and accurately analyzing the input data, AI-CRM could bring success to the organizations.

10 Conclusion

With the help of accurate, curated, and actionable data of customers, this study has demonstrated that applications of AI-CRM will derive considerable benefits to the concerned organizations. A business organization must have the ability of possessing accurate expertise and staffing helpful to mine, capture, and synthesize the collected data to predict right behavior of customers by taking help of AI technology. This is what is known as AI-integrated CRM capability. All the successful organizations would like to have this tool (Thrassou et al., 2021c). An effective and comprehensive AI-CRM platform will be able to provide the employees easy access to such necessary data, and it will help the customers towards taking accurate purchase decision. Thus, CRM activities integrated with AI will help the organizations to improve the relationships between organizations and potential customers which in turn would help the organizations to achieve greater success (Rana et al., 2021b). Thus, it is seen that an organizational performance can be improved by adopting AI-integrated CRM system helping the organizations towards real-time decision-making.

References

Akter, S., Bandara, R., Hani, U., Wamba, S. F., Foropon, C., & Papadopoulos, T. (2019). Analytics-based decision-making for service systems: A qualitative study and agenda for future research. *International Journal of Information Management, 48*(3), 85–95.

Aleksej, H., Marie, G., Alex, F., & Gordon, F. (2018). Knowledge exchange partnership leads to digital transformation at Hydro-X Water Treatment, Ltd. *Global Business and Organizational Excellence, 37*(4), 6–13.

Alshare, K. A., & Lane, P. L. (2011). Predicting student-perceived learning outcomes and satisfaction in ERP courses: An empirical investigation. *Communications of the Association for Information Systems, 28*(1), 571–584.

Alshibly, H. H. (2015). Customer perceived value in social commerce: An exploration of its antecedents and consequences. *Journal of Management Research, 7*(1), 17–37.

Awasthi, P., & Sangle, P. S. (2013). The importance of value and context for mobile CRM services in banking. *Business Process Management Journal, 19*(6), 864–891.

Baabdullah, A. M., Chatterjee, S., Rana, N., & Dwivedi, Y. K. (2021). Understanding AI adoption in manufacturing and production firms using an integrated TAM-TOE model, *170*(7), 120880.

Bag, S., Gupta, S., Kumar, A., & Sivarajah, U. (2021). An integrated artificial intelligence framework for knowledge creation and B2B marketing rational decision making for improving firm performance. *Industrial Marketing Management, 92*, 178–189.

Balakrishnan, J., & Dwivedi, Y. K. (2021). Role of cognitive absorption in building user trust and experience. *Psychology & Marketing, 38*(4), 643–668.

Basile, G., Chatterjee, S., Chaudhuri, R., & Vrontis, D. (2021). Digital transformation and entrepreneurship process in SMEs of India: A moderating role of adoption of AI-CRM capability and strategic planning. *Journal of Strategy and Management.* https://doi.org/10.1108/JSMA-02-2021-0049

Bhattacharya, K., & Chatterjee, S. (2020). Adoption of artificial intelligence in higher education: A quantitative analysis using structural equation modelling. *Education and Information Technologies, 24*(1), 3443–3463.

Bishop, D. (2009). Why existing customers are so valuable. *Strategic Direction, 25*(2), 3–5.

Bose, R. (2002). Customer relationships management: Key components for IT success. *Industrial Management and Data Systems, 102*(2), 89–97.

Bounfour, A. (2016). *Digital futures, digital transformation* (pp. 55–73). Springer.

Bowen, J. P., & Giannini, T. (2014). *Digitalism: The New Realism?* EVA London 2014Conference Proceedings. Electronic Workshops in Computing (eWiC). BCS, pp. 324–331. https://doi.org/10.14236/ewic/eva2014.38.

Bradlow, E. T., Gangwar, M., Kopalle, P., & Voleti, S. (2017). The role of big data and predictive analytics in retailing. *Journal of Retailing, 93*(1), 79–95.

Brynjolfsson, E., & Mcafee, A. (2017). The business of artificial intelligence. *Harvard Business Review.*

Cannon, J. P., & Perreault, W. D. (1999). Buyer-seller relationship in business markets. *Journal of Marketing Research, 36*(4), 439–460.

Chatterjee, S. (2019a). Impact of AI regulation on intention to use robots: From citizens and government perspective. *International Journal of Intelligent Unmanned Systems, 8*(2), 97–114.

Chatterjee, S. (2019b). Influence of IoT policy on quality of life: From government and Citizens' perspective. *International Journal of Electronic Government Research, 15*(2), 19–38.

Chatterjee, S. (2019c). Is data privacy a fundamental right in India? An analysis and recommendations from policy and legal perspective. *International Journal of Law and Management, 61*(1), 170–190.

Chatterjee, S. (2019d). Why people will use IoT enabled devices? An empirical examination from Indian perspectives. *International Journal of Technology and Human Interaction, 18*(2), 33–61.

Chatterjee, S. (2020). AI strategy of India: Policy framework, adoption challenges and actions for government. *Transforming Government: People, Process and Policy, 14*, 757. https://doi.org/10.1108/TG-05-2019-0031

Chatterjee, S. (2021a). Antecedence of attitude towards IoT usage: A proposed unified model for IT professionals and its validation. *International Journal of Human Capital and Information Technology Professionals, 12*(2), 13–34.

Chatterjee, S. (2021b). Impact of addiction of online platforms on quality of life: Age and gender as moderators. *Australasian Journal of Information Systems., 25*(1), 102–134.

Chaudhuri, R., Chatterjee, S., & Vrontis, D. (2021). Examining the global retail apocalypse during the COVID-19 pandemic using strategic omnichannel management: A consumers' data privacy and data security perspective. *Journal of Strategic Marketing.* https://doi.org/10.1080/0965254X.2021.1936132

Cochoy, H., & Peterson McIntyre, S. (2017). *Digitalizing consumption.* Routledge Publication.

Culotta, A., Ravi, N. K., & Cutler, J. (2015). Predicting the demographics of twitter users from website traffic data. In 29th AAAI conference on Artificial Intelligence. AAAI 2015 and the 27th Innovative Applications of Artificial Intelligence Conference.

Damaševičius, R., Chatterjee, S., Majumdar, D., & Misra, S. (2020). Adoption of mobile applications for teaching-learning process in rural girls' schools in India: An empirical study. *Education and Information Technologies, 25*, 4057–4076.

Davenport, T., Guha, A., Grewal, D., & Bressgott, T. (2019). How artificial intelligence will change the future of marketing. *Journal of the Academy of Marketing Science, 48*(1), 24–42.

Dwivedi, Y. K., Chatterjee, S., & Kar, A. K. (2021). Intention to use IoT by aged Indian consumers. *Journal of Computer Information Systems.* https://doi.org/10.1080/08874417.2021.1873080

Fountaine, T., McCarthy, B., & Saleh, T. (2019). Building the AI-powered organization. *Harvard Business Review*, 63–73.

Gantz, J.F., Gerry, M., David, S., Dan, V., & Mary, W. (2017). *A trillion-Dollar boost: The economic impact of AI on customer relationship management.* Sales force publication, pp. 1–20. Available at: https://www.salesforce.com/content/dam/web/en_us/www/documents/white-papers/the-economic-impact-of-ai.pdf. Accessed on 15-07-2021.

Ghosh, S. K., Chatterjee, S., & Chaudhuri, R. (2019). Adoption of ubiquitous customer relationship management (uCRM) in Enterprise: Leadership support and technological competence as moderators. *Journal of Relationship Marketing, 19*(2), 75–92.

Ghosh, S. K., Chatterjee, S., & Chaudhuri, R. (2020). Knowledge management in improving business process: An interpretative framework for successful implementation of AI–CRM–KM system in organizations. *Business Process Management Journal, 26*(6), 1261–1281.

Ghosh, S. K., Chatterjee, S., Chaudhuri, R., & Chaudhuri, S. (2021). Adoption of AI-integrated CRM system by Indian industry: From security and privacy perspective. *Information and Computer Security, 29*(1), 1–24.

Gobble, M. A. M. (2018). Digitalization, digitization, and innovation research-technology management. *Taylor &Francis., 61*(4), 56–59.

Graca, S. S., Barry, J. M., & Doney, P. M. (2015). Performance outcomes of behavioral attributes in buyer-supplier relationships. *Journal of Business & Industrial Marketing, 30*(7), 805–816.

Grönroos, C. (1990). *Service management and marketing: Managing the moments of truth in service competition.* Jossey-Bass.

Grönroos, C. (1994). From marketing mix to relationship marketing: Towards a paradigm shift in marketing. *Management Decision, 32*(2), 4–20.

Grover, P., Kar, A. K., & Dwivedi, Y. K. (2020). Understanding artificial intelligence adoption in operations management: Insights from the review of academic literature and social media discussions. *Annals of Operations Research, 308*, 177. https://doi.org/10.1007/s10479-020-03683-9

Gupta, S., Leszkiewicz, A., Kumar, V., Bijmolt, T., & Potapov, D. (2020). Digital analytics: Modeling for insights and new methods. *Journal of Interactive Marketing, 51*(8), 26–43.

Harmeling, C. M., Palmatier, R. W., Houston, M. B., Arnold, M. J., & Samaha, S. A. (2015). Transformational relationship events. *Journal of Marketing, 79*(5), 39–62.

Härting, R.-C., Reichstein, C., & Jozinovic, P. (2017). The potential value of digitization for business. *Informatik, 275*, 647–1657.

Heide, J. B., Wathne, K. H., & Rokkan, A. I. (2007). Interfirm monitoring, social contracts, and relationship outcomes. *Journal of Marketing Research, 44*(3), 425–433.

Hogg, G., & Gabbott, M. (1998). *Consumers and services.* Wiley. ISBN 0 471 96269 4. 271. https://doi.org/10.1002/cb.75

Hopkinson, P. J., Perez-Vega, R., & Singhal, A. (2018). *Exploring the use of AI to manage customers' relationships.* Conference: Academy of Marketing Workshop: Artificial Intelligence in Marketing – The field, research directions, and methodological issues, pp. 1–8.

Hu, Q., Lu, Y., Pan, Z., Gong, Y., & Yang, Z. (2021). Can AI artifacts influence human cognition? The effects of artificial autonomy in intelligent personal assistants. *International Journal of Information Management, 56*(7), 102250.

Jabbar, A., Akhtar, P., & Dani, S. (2019). Real-time big data processing for instantaneous marketing decisions: A problematization approach. *Industrial Marketing Management., 90*(10), 558–569.

Jane M. C. (2015). *The company cultures that help (or hinder) digital transformation.* Available at: https://hbr.org/2015/08/the-company-cultures-that-help-or-hinder-digital-transformation. Harvard Business Review. Accessed on 22-07-2021.

Kar, A. K., & Chatterjee, S. (2020). Why do small and medium enterprises use social media marketing and what is the impact: Empirical insights from India. *International Journal of Information Management, 53*(8), 102013.

Kennedy, A. (2006). Electronic customer relationship management (eCRM): Opportunities and challenges. *Irish Marketing Review, 18*(1/2), 58–69.

Khan, S. (2017). *Leadership in the digital age – A study on the effects of digitalization on top management leadership.* Thesis. Stockholm Business School. Available at: https://su.diva-portal.org/smash/get/diva2:971518/FULLTEXT02.pdf. Accessed on 22 July 2021.

Kumar, V., Dixit, A., Javalgi, R., & Dass, M. (2016). Research framework, strategies, and applications of intelligent agent technologies (IATs) in marketing. *Journal of the Academy of Marketing Science, 44*(1), 24–45.

Li, M., & Nguyen, B. (2016). When will firms share information and collaborate to achieve innovation? A review of collaboration strategies. *The Bottom Line, 30*(1), 65–86.

Libai, B., Bart, Y., Gensler, S., Hofacker, C. F., Kaplan, A., Kötterheinrich, K., & Kroll, E. B. (2020). Brave new world? On AI and the management of customer relationships. *Journal of Interactive Marketing, 51*(2), 44–56.

Maxwell, A. L., Jeffrey, S. A., & Lévesque, M. (2011). Business angel early-stage decision making. *Journal of Business Venturing, 26*(2), 212–225.

Mikalef, P., & Gupta, M. (2021). Artificial intelligence capability: Conceptualization, measurement calibration, and empirical study on its impact on organizational creativity and firm performance. *Information & Management, 58*(3), 103434.

Molinillo, S., & Japutra, A. (2017). Organizational adoption of digital information and technology: A theoretical review. *The Bottom Line, 30*(1), 33–46.

Nguyen, B., & Chatterjee, S. (2021). Value co-creation and social media at bottom of pyramid (BOP). *The Bottom Line, 34*, 101. https://doi.org/10.1108/BL-11-2020-0070

Nguyen, B., Chatterjee, S., Ghosh, S., & Chaudhuri, R. (2019). Are CRM systems ready for AI integration? *The Bottom Line, 32*(2), 144–157.

Nguyen, B., Chatterjee, S., Ghosh, S. K., Bhattacharjee, K. K., & Chaudhuri, S. (2020). Adoption of artificial intelligence integrated CRM system: An empirical study of Indian organizations. *The Bottom Line, 33*(4), 359–375.

Nishant, R., Kennedy, M., & Corbett, J. (2020). Artificial intelligence for sustainability: Challenges, opportunities, and a research agenda. *International Journal of Information Management, 53*(2), 102104.

O'Donnell, J. (2017). *IDC says get on board with the DX economy or be left behind.* https://searcherp.techtarget.com/news/450414723/IDC-says-get-on-board-with-the-DX-economy-or-beleft-behind.techtarget.com

Ostrom, A. L., Parasuraman, A., Bowen, D. E., Patricio, L., & Voss, C. A. (2015). Service research priorities in a rapidly changing context. *Journal of Service Research, 18*(2), 127–159.

Padmanabhan, B., & Tuzhilin, A. (1999). Unexpectedness as a measure of interestingness in knowledge discovery. *Decision Support Systems, 27*(3), 303–318.

Parise, S., Guinan, P. J., & Kafka, R. (2016). Solving the crisis of immediacy: How digital technology can transform the customer experience. *Business Horizons, 59*(4), 411–420.

Paulus-Rohmer, D., Schatton, H., & Bauernhansl, T. (2016). Ecosystems, strategy and business models in the age of digitization - how the manufacturing industry is going to change its logic. *Procedia CIRP, 57*, 8–13.

Peelen, E. (2005). *Customer relationship management. Edinburgh gate* (pp. 11–12). Pearson Education Limited.

Piccolo, R., Chatterjee, S., Chaudhuri, R., & Vrontis, D. (2021). Enterprise social network for knowledge sharing in MNCs: Examining the role of knowledge contributors and knowledge seekers for cross-country collaboration. *Journal of International Management, 27*(1), 100827.

Raghunathan, S. (1999). Impact of information quality and decision-maker quality on decision quality: A theoretical model and simulation analysis. *Decision Support Systems, 26*(4), 275–286.

Rana, N. P., Chatterjee, S., & Dwivedi, Y. K. (2020). Social media as a tool of knowledge sharing in academia: An empirical study using valance, instrumentality and expectancy (VIE) approach. *Journal of Knowledge Management, 24*(10), 2531–2552.

Rana, N. P., Chatterjee, S., & Dwivedi, Y. K. (2021a). Assessing consumers' co-production and future participation on value co-creation and business benefit: An F-P-C-B model perspective. *Information Systems Frontiers, 24*, 945. https://doi.org/10.1007/s10796-021-10104-0

Rana, N. P., Chatterjee, S., & Dwivedi, Y. K. (2021b). How does business analytics contribute to organisational performance and business value? A resource-based view. *Information Technology & People.* https://doi.org/10.1108/ITP-08-2020-0603. ISSN: 0959-3845.

Reinartz, W., Wiegand, N., & Imschloss, M. (2018). The impact of digital transformation on the retailing value chain. *International Journal of Research in Marketing, 36*(3), 350–366.

Rust, R. Y. (2020). The future of marketing. *International Journal of Research in Marketing, 37*(1), 15–26.

Rust, R. T., & Huang, M.-H. (2014). The service revolution and the transformation of marketing science. *Marketing Science, 33*(2), 206–221.

Sahu, G., & Gupta, M. P. (2007). Users' acceptance of e-government: A study of Indian central excise. *International Journal of Electronic Government Research, 3*(3), 1–21.

Sakka, G., Chatterjee, S., Chaudhuri, R., & Thrassou, A. (2021a). Impact of firm's intellectual capital on firm performance: A study of Indian firms and the moderating effects of age and gender. *Journal of Intellectual Capital, 23*, 103. https://doi.org/10.1108/JIC-12-2020-0378

Sakka, G., Chatterjee, S., Chaudhuri, S., & Apoorva, A. (2021b). Cross-disciplinary issues in international marketing: A systematic literature review on international marketing and ethical issues. *International Marketing Review, 38*, 985. https://doi.org/10.1108/IMR-12-2020-0280

San-Martína, S., Jiméneza, N. H., Lopez, C., & B. (2016). The firms benefit of mobile CRM from the relationship marketing approach and the TOE model. *Spanish Journal of Marketing – ESIC, 20*(1), 18–29.

Saura, J. R., Palos-Sanchez, P., & Blanco-González, A. (2019). The importance of information service offerings of collaborative CRMs on decision-making in B2B marketing. *Journal of Business & Industrial Marketing, 35*(3), 470–482.

Schultz, M., & Pick, D. (2012). From CM to CRM to CN2: A research agenda for the marketing communications transition. *Advances in Advertising Research, 3*, 421–432.

Siachou, E., Chatterjee, S., Chaudhuri, R., & Vrontis, D. (2021). Examining the dark side of human resource analytics: An empirical investigation using the privacy calculus approach. *International Journal of Manpower*. https://doi.org/10.1108/IJM-02-2021-0087

Sreenivasulu, N. S., & Chatterjee, S. (2019). Personal data sharing and legal issues of human rights in the era of artificial intelligence: Moderating effect of government regulation. *International Journal of Electronic Government Research, 15*(3), 21–36.

Sreenivasulu, N. S., & Chatterjee, S. (2021). Artificial intelligence and human rights: A comprehensive study from Indian legal and policy perspective. *International Journal of Law and Management*. https://doi.org/10.1108/IJLMA-02-2021-0049

Syam, N., & Sharma, A. (2018). Waiting for a sales renaissance in the fourth industrial revolution: Machine learning and artificial intelligence in sales research and practice. *Industrial Marketing Management., 69*(2), 135–146.

Thrassou, A., Chatterjee, S., Chaudhuri, R., & Vrontis, D. (2021a). The influence of online customer reviews on customers' purchase intentions: A cross-cultural study from India and the UK. *International Journal of Organizational Analysis*. https://doi.org/10.1108/IJOA-02-2021-2627

Thrassou, A., Chatterjee, S., Chaudhuri, R., & Vrontis, D. (2021b). Does "CHALTA HAI" culture negatively impacts sustainability of business firms in India? An empirical investigation. *Journal of Asia Business Studies*. https://doi.org/10.1108/JABS-12-2020-0471

Thrassou, A., Chatterjee, S., Chaudhuri, R., & Vrontis, D. (2021c). Antecedents and consequences of knowledge hiding: The moderating role of knowledge hiders and knowledge seekers in organizations. *Journal of Business Research., 128*(6), 303–313.

Tsai, C. W., Chatterjee, S., Bhattacharjee, K. K., & Agrawal, A. K. (2021). Impact of peer influence and government support for successful adoption of technology for vocational education: A quantitative study using PLS-SEM technique. *Journal of Quality and Quantity, 55*, 2041. https://doi.org/10.1007/s11135-021-01100-2

Verma, D., & Verma, D. S. (2013). Managing customer relationships through mobile CRM in organized retail outlets. *International Journal of Engineering Trends and Technology, 4*(5), 1696–1701.

Vogelsang, M. (2010). *Digitalization in open economies, contributions to economics*. Physica-Verlag Publication.

Vrontis, D., Chatterjee, S., & Chaudhuri, R. (2020). Article title: Does data-driven culture impact innovation and performance of a firm? An empirical examination. *Annals of Operational Research*, 1. https://doi.org/10.1007/s10479-020-03887-z

Vrontis, D., Chatterjee, S., & Chaudhuri, R. (2021a). Knowledge sharing in international markets for product and process innovation: Moderating role of firm's absorptive capacity. *International Marketing Review, 39*, 706. https://doi.org/10.1108/IMR-11-2020-0261

Vrontis, D., Chatterjee, S., Chaudhuri, R., Thrassou, A., & Ghosh, S. K. (2021b). ICT-enabled CRM system adoption: A dual Indian qualitative case study and conceptual framework development. *Journal of Asia Business Studies, 15*(2), 257–277.

Vrontis, D., Chatterjee, S., Chaudhuri, R., Thrassou, A., Ghosh, S. K., & Chaudhuri, S. (2021c). Social customer relationship management factors and business benefits. *International Journal of Organizational Analysis, 29*(1), 35–58.

Wang, Y., Lewis, M., Cryder, C., & Sprigg, J. (2016). Enduring effects of goal achievement and failure within customer loyalty programs: A large-scale field experiment. *Marketing Science, 35*(4), 565–575.

Watson, G. F., Beck, J. T., Henderson, C. M., & Palmatier, R. W. (2015). Building, measuring, and profiting from customer loyalty. *Journal of the Academy of Marketing Science, 43*(6), 790–825. https://doi.org/10.1007/s11747-015-0439-4

Wen, K.-W., & Chen, Y. (2010). E-business value creation in small and medium enterprises: A US study using the TOE framework. *International Journal of Electronic Business, 8*(1), 80–100.

Wierenga, A. (2009). Young people making a life. *Sage Book Publication, 17*(4), 457–458.

Efficient Supplier Selection in the Era of Industry 4.0

Deepanshu Nayak, Meenu Singh, Millie Pant, and Sunil Kumar Jauhar

1 Introduction

Industries play an important role in the growth and development of every country. Industries are growing continuously along with the passage of time (Yin et al., 2018). The different generations depicted in Fig. 1 represent the evolution of industries. The first generation (from the eighteenth to nineteenth century) featured hand-operated devices and steam engines. Manufacturing volume and product accuracy were low during this time period. After 1870, commercial distribution of electricity began in the second generation, resulting in the use of an assembly line in the manufacturing system and the beginning of mass production. During the Third Industrial Revolution, programmable chips, computers, and robotics surged and increased after 1969. As a result of applying these technologies, manpower was reduced, and the manufacturing system became automatic. Additionally, the concept of supply chain management (SCM) for the efficient flow of information and material was proposed in this era. SCM is defined as *designing and management of all activities involved in procurement, production, inventory, transportation, and information* (Janvier-James, 2012)

D. Nayak (✉) · M. Singh
Department of Applied Science and Engineering, IIT Roorkee, Roorkee, Uttarakhand, India
e-mail: dnayak@pp.iitr.ac.in; msingh1@as.iitr.ac.in

M. Pant
Department of Applied Science and Engineering, IIT Roorkee, Roorkee, Uttarakhand, India

Centre for Artificial Intelligent and Data Science, IIT Roorkee, Roorkee, Uttarakhand, India
e-mail: millifpt@iitr.ac.in

S. K. Jauhar
Department of Operations and Decision Sciences, IIM Kashipur, Kashipur, Uttarakhand, India
e-mail: sunil.jauhar@iimashipur.ac.in

© The Author(s), under exclusive license to Springer Nature Switzerland AG 2023
S. S. Kamble et al. (eds.), *Digital Transformation and Industry 4.0 for Sustainable Supply Chain Performance*, EAI/Springer Innovations in Communication and Computing, https://doi.org/10.1007/978-3-031-19711-6_9

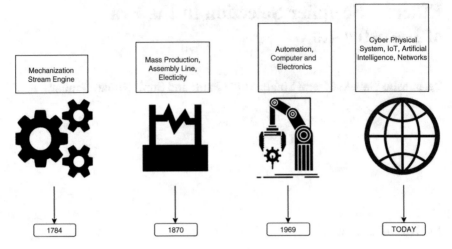

Fig. 1 Evolution of industry

As the present era is characterized by digitalization and technological growth, industries are transforming themselves into a digital environment; this revolution of the transformation by the industries is known as Industry 4.0. In this era, industries integrate their operations with CC, AI (artificial intelligence), IoT, ICT, and other technologies (Xu et al., 2018).

The procurement process is an important part of supply chain for any industry. This process determines the quality of raw materials, the life cycle of the product, the rate of delivery, and other parameters. In the age of Industry 4.0, not only production processes should be smart, but SCM should also be digitalized (Ghadimi et al., 2019). Choosing a suitable supplier is a critical decision for a successful procurement process. Multi-criteria decision-making (MCDM) approaches are extensively utilized for the selection of appropriate supplier as these procedures are applicable in both crisp and fuzzy environments (Çalık, 2021).

The aim of this study is to provide an overview and comparison of the conventional methods/models available in literature with the role of Industry 4.0. In Sect. 2 an overview of literature review is presented related to supplier selection in the era of Industry 4.0. The past studies along with the field of application, criteria in supplier selection considered are represented in Sect. 3. Selected models are represented in Sect. 4 of the present study. Finally in Sect. 4, conclusions based on the present study are provided.

2 Literature Review

In the recent years, limited research on supplier selection for the Industry 4.0 era have been conducted. This section presents an overview of hybrid MCDM techniques used for supplier selection. This section demonstrates the wide applications of fuzzy approaches such as Pythagorean fuzzy numbers, FAHP (fuzzy analytic hierarchy process), FTOPSIS (fuzzy technique for order of preference by similarity to ideal solution), DEMATEL (decision-making trial and evaluation laboratory), and COPRAS-G (complex proportional assessment of alternatives with gray relations). MAS (multi-agent system) and fuzzy-neuro approaches are employed for the data collection, storage and analysis, etc. Digital criteria selected for supplier selection are introduced and explained in Sect. 2.1.

2.1 Description of Criteria/Features

Criteria make the key difference between the conventional studies of supplier selection and the present digital study. Criteria are factors that have an impact on the objective function, either directly or indirectly. They can be identified through a review of the literature and discussions with decision-makers. Criteria may be qualitative or quantitative in nature (Singh & Pant, 2021). Qualitative criteria are represented by linguistic terms like good, very good, high, low, etc. Quantitative criteria, on the other hand, have a specific value. Beneficial and non-beneficial criteria are another classification of criteria. Beneficial criteria are those that should be maximized and have a positive impact on the objective function. Non-beneficial criteria are also known as cost criteria since they have a negative impact on the objective function; hence, the value of these criteria should be reduced during optimization, i.e., minimized.

In the era of Industry 4.0, supplier selection criteria are collaborated with the conventional study criteria. Table 1 discusses certain criteria related to Industry 4.0, commonly known as digital criteria.

Many studies have been performed utilizing multi-criteria decision-making (MCDM) approaches for the supplier selection in forward or reverse SCM. Tables 2 and 3 show the previous studies in forward and reverse SC, highlighting their purpose as well as the criteria chosen for that study.

Table 1 Definition of criteria used for supplier selection in Industry 4.0

Criteria	Description	References
Internet of things (IoT)	IoT tools are used to analyze the real-time movement/transportation of commodities and products along the supply chain. This technology provides fast and accurate information about suppliers and manufacturers.	Çalık (2021), Sachdeva et al. (2019), Kaya and Aycin (2021)
Big data analytics and cloud computing	To collect data and apply the smart devices and big data analytics in the supply chain management to improve the service.	Çalık (2021), Kaya and Aycin (2021)
Automation	This includes CIM (computer-integrated manufacturing), assembly line with minimal operators, and automatic feedback of output, smart PPC (production planning and control). It decreases the cycle time of the product and increases the efficiency of production.	Çalık (2021)
Digital collaboration	This reveals that the manufacturers and suppliers are connected by a digital network like the internet and share their data digitally.	Zekhnini et al. (2020), Büyüközkan and Göçer (2019)
Cybersecurity	It is the protection of digital data and information from any theft. Otherwise, the shared information and data may be destroyed or changed by which results obtained would be inaccurate.	Büyüközkan and Göçer (2019), Hasan et al. (2020)
ICT (information and communications technology)	This technology enables the industry to form the infrastructure of modern computing.	Büyüközkan and Göçer (2019), ÖZBEK and Yildiz (2020)
Robotics	Many tasks, including material handling, processing, feeding, drilling, and so on, may now be carried out automatically with the help of robots.	Çalık (2021); Özbek and Yildiz (2020)
Smart logistics	To enable logistics operations like inventory management, planning, and transportation with modern digital technology.	Kaya and Aycin (2021)

3 Supplier Selection Methods in the Era of Industry 4.0

Industry 4.0 is relatively a new concept, and the works related to supplier selection in the era of Industry 4.0 have only come into focus in the last 3–5 years. From these works, only the most significant ones are selected and are discussed in parts in the following sections.

Table 2 Past studies on forward supplier selection in Industry 4.0

Year	Objective	Approaches	Criteria	References
2017	Selection of suppliers among SMEs using BWM and fuzzy TOPSIS based on their green innovation ability	Small and medium enterprises	Environmental investments and economic benefits, collaborations, environmental management initiatives, green purchasing capabilities, regulatory obligations, pressures and market demand, availability of resources and green competencies, research and design initiatives	Gupta and Barua (2017)
2017	Selection of material using MCDM methods like TOPSIS, DEA, and COPRAS	Material selection	Specific heat, Young's modulus, yield strength, density, thermal conductivity, thermal expansion index	Mousavi-Nasab and Sotoudeh-Anvari (2017)
2018	Selection of intelligent supplier	Multi-agent systems (MASs), FIS model with triangular membership function	Environmental dimension: Green image, pollution control, green competencies	Ghadimi et al. (2019)
			Economic dimension: Quality, service/delivery, cost, technical capability	
			Social dimension	
2018	Selection of a supplier for the development of a new product using the MCDM approach	New product development (NPD) environment	Technological compatibility, social acceptability, economy, environment, quality	Sinha and Anand (2018)
2019	Supplier selection in the era of industry 4.0	Intuitionistic fuzzy weighted approach, fuzzy entropy weight-based multi-criteria decision model with TOPSIS	Delivery delay, cost/price, industry 4.0 tech-enabled, relationship, rejection rate	Sachdeva et al. (2019)
2019	Sustainable supplier selection based on AHPSort II in interval type-2 fuzzy environment	Supplier selection of M company	Economic: Price, profit, transportation cost	Xu et al. (2019)
			Environmental: Energy conservation, pollution control	
			Social: Information disclosure, safety audit, and assessment	

(continued)

Table 2 (continued)

Year	Objective	Approaches	Criteria	References
2019	Selection of the sustainable supplier selection in importing field with an integrating ANP and VIKOR method	Ranking of the city for their manufacturing process	Economic factor: Cost of transportation, cost of product, product revenue	Abdel-Baset et al. (2019)
			Environmental factor: Labelling and green packing, green manufacturing, trash management	
			Social factor: Legal compliance and ethical issues, safety and health system, information revelation	
2019	Selection of a sustainable supplier using the fuzzy TOPSIS method	Automated casting system	Economic: Quality of products, cost, service performance	Memari et al. (2019)
			Environmental sustainability: Green image, environmental efficiency, green competencies, pollution reduction	
			Social sustainability: Employment practices, health and safety	
2020	Selection of sustainable supplier selection using MACROS in the healthcare industry	Healthcare industry	Economic criteria: Reliability, price, innovativeness, on-time delivery, organization and management, assortment width	Stević et al. (2020)
			Social criteria: Information disclosure, security and disciplinary practices, reputation, respect for the policies, labor health and work safety, training	
2020	Selection of green supplier	Pythagorean fuzzy numbers, PFAHP, PFTOPSIS	Delivery: Automated guided vehicles, robotics, service level	Çalık (2021)
			Pollution control: Lean automation, process safety, and environmental control	
			Production: IoT and CPS, cloud computing, big data analytics	
			Quality: Quality 4.0, augmented reality and 3D printing	
			Environmental representation: Green design, green image	

Year	Objective	Method	Criteria/description	Reference
2020	To represent the model for selection of intelligent supplier	Digital supply chain, resilience, sustainability, adaptive fuzzy-neuro approach	Primary performance criteria: Cost, quality, delivery	Zekhnini et al. (2020)
			Supplier's technological capability: Flexibility, reliability, responsiveness, visibility, real-time, agility	
			Suppliers resilience: Ambidextry, vulnerability, collaboration, risk awareness, viability	
			Supplier's sustainability: Environmental competencies, social competencies, economic competencies	
2020	Selection of sustainable and smart supplier	DMATEL, TOPSIS	Economic practices: Improving the quality of the product based on big data analytics, by smart technology reduction in the cost, enhancement of supply flexibility, smart delivery of the product	Chen et al. (2020)
			Environmental practices: Based on a digital platform green, digitally purchasing green design, using smart technologies management of internal awareness for enhancing green development, smart and green manufacturing, smart and logistics	
			Social practices: Confirm the rights of partners using smart technology of perception, to promote the smart technology using social activities, smart working environment	
2021	To select the right suppliers in the era of industry 4.0 by identifying key criteria and evaluate them	AHP, COPRAS-G, interval type-2 fuzzy set	RFID and dynamic sensors, quality, cost/price, delivery, capacity, intelligent transportation systems such as GPS, smart warehouse and shelving system, employee training on industry 4.0, use of an autonomous machine, big data and cloud computing, internet of things (IoT) implementation	Kaya and Aycin (2021)

Table 3 Past studies on reverse supplier selection in Industry 4.0

Year	Objective	Industry/Area	Criteria	References
2014	To select the reverse logistics supplier for the Indian mobile industry	Mobile industry	Availability of skilled personnel, impacts of environmental pollution, availability of a covered and closed area, inspection/sorting and disassembly cost, mobile recycling cost, level of noise pollution, possibilities to work with NGOs, mobile phone refurbishing cost, e-waste storage capacity, safe disposal cost	Jayant et al. (2014)
2014	A robust hybrid MCDM methodology for contractor evaluation and selection in third-party reverse logistics	Plastic industry (PET)	Reverse logistics process functions, organizational performance criteria, organizational role of RL, resources capacity, enterprise alliance, quality of service, communication systems, experience, location	Senthil et al. (2014)
2015	To overcome reverse logistics adoption barriers under fuzzy environment using the combination of AHP-TOPSIS method for prioritizing the solutions	Electronic industry	Organizational barriers, management barriers, economic barriers, market-related barriers, legal barriers, infrastructural barriers, technological barriers	Prakash and Barua (2015)
2016	To select the third-party reverse logistics supplier selection using combined MCDM approach	Electronics organization	Strategic barriers, organizational barriers, economical barriers, policy barriers, technological barriers, infrastructural barriers, marketing barriers	Prakash and Barua (2016)
2016	To represent an effective model for remanufacturing process and supplier selection in a supply chain having the close loop	Watch manufacturing	Delivery time, cost, defect rate, technical knowledge, reputation	Shakourloo et al. (2016)
2018	Using cumulative prospect theory and hybrid-information MCDM selection of third-party reverse logistics provider approach	PCs industry	Reverse logistics cost, reverse logistics revenue, organizational functions, quality of service, company capacity/ competence, strategic alliance, environmental friendliness	Li et al. (2018)

3.1 Selection of Intelligent Suppliers by Multi-agent Theory (Ghadimi et al., 2019)

This study was proposed by P. Ghadimi in 2018. According to this study, the selection of supplier in the SC, physical interactions between the members of SC are reduced by using the tools of Industry 4.0. In this case, MAS (multi-agent system) is used to select the intelligent suppliers. This system is useful for receiving real-time information in a transparent manner, decentralizing SC members, establishing a structured communication channel, and so on.

MAS is composed of three layers, namely, interface, technical, and data resources. Figure 2 depicts the structure of the MAS system. The agent in the first layer is directly connected to the user and collects information or data. Supplier databases, supplier knowledge databases, and manufacturer databases are examples of data resource layers. Supplier databases contain information about all suppliers, whereas manufacturer databases contain information about the product and its components. Three agents are designed for supplier evaluation in the technological layer: DBA (database agent), DMA (data maker agent), and SA (supplier agent).

In this case, SA acts as a user agent, receiving supplier input data. This input data is sent by SA to the DBA, who send confirmation to SA after receiving the input data. Following that, SA requests DMA to evaluate the results, and after the evaluation, SA receives the performance of suppliers from DMA. The DBA stores the input data received from the SA. DMA requests input data from DBA. DBA sends the saved input data to the DMA for evaluation. The evaluation results are also saved in the database by the DBA. This system provides real-time data and reduces the degree of uncertainty in supplier evaluation (Ghadimi et al., 2019).

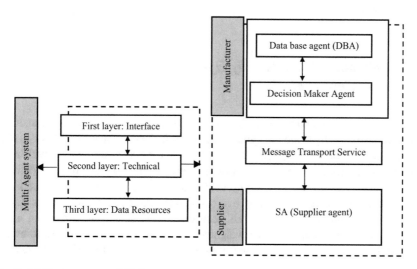

Fig. 2 Multi-agent system (MAS) system

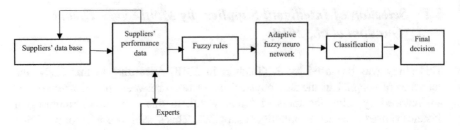

Fig. 3 An adaptive fuzzy-neuro approach

3.2 Selection of the Suppliers Through Adaptive Fuzzy-Neuro Approach (Zekhnini et al., 2020)

This study is performed by K. Zekhnini in 2020 (Zekhnini et al., 2020). The objective of this study is to select sustainable supplier selection in uncertain conditions. The criteria used for the sustainable study are economic (all types of cost, profits), social (relationship with customers, employee training), and environmental criteria. The structure of this study is depicted in Fig. 3.

To study the smart supply chain, some additional criteria like technological capability of Industry 4.0, smartness, and resilience are used. The author employs an adaptive fuzzy-neuro approach in this study. This approach is made up of six layers. The value of criteria is entered in the first layer. The second layer converts the given values of criteria into the fuzzy number. The third layer is the rule layer which applies the approach on the input by given the experts. The fourth layer evaluates every rule, after that in the next layer, defuzzification is done, and, finally, the last layer provides the output in terms of the ranking score of the suppliers (Zekhnini et al., 2020).

3.3 Selection of Suppliers by Hybrid MCDM Approach (Sachdeva et al., 2019)

This study is represented by N. Sachdeva in 2021 for the selection of suppliers in the era of Industry 4.0 by using the hybrid multi-criteria decision-making (MCDM) tool. In this study, Industry 4.0 technology-enabled is used as criteria along with more conventional factors such as cost, pricing, relationship (social criteria), delivery delay, and rejection rate. The supply chain becomes digital once these criteria are met. Fuzzy set theory is used to overcome the condition of uncertainty during supplier selection. In this study, all decision-makers (DM) are not given equal weight; hence, an intuitive fuzzy weighted approach (IFWA) is utilized to assign weights to them. With the help of decision-makers, comparison matrix is formulated, and with the help of the entropy method, weight of the criteria is calculated. Following that, a weighted matrix of different suppliers is created, and the rank of alternatives is calculated using the TOPSIS method (Sachdeva et al., 2019).

3.4 Selection of Suppliers by Hybrid Fuzzy MCDM Approach (Kaya & Aycin, 2021)

This study is represented by SK Kaya in 2021 to determine the key criteria associated with the Industry 4.0 era. The steps involved in the study are depicted in Fig. 4. The main aim of the study is to analyze the effects of Industry 4.0 criteria on the selection of an efficient supplier. A case study of supplier selection for the textile manufacturer is carried out. This study comprises conventional criteria like cost/price, capacity, delivery, and some criteria associated with Industry 4.0 that are

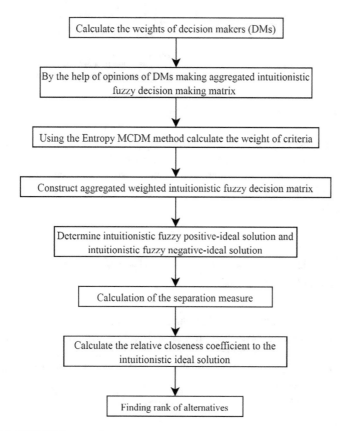

Fig. 4 Hybrid MCDM approach

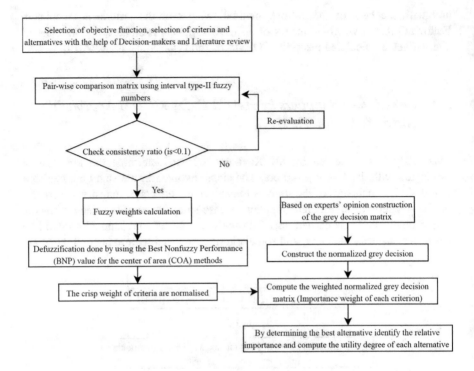

Fig. 5 An integrated interval type-2 fuzzy AHP and COPRAS-G approach

selected from the literature review. After applying the fuzzy AHP method, the consistency ratio is checked whether it is less than 10% or not. If yes, the weight of the criteria is calculated, which are further used to calculate the rank of the different available suppliers by using the COPRAS-G method (Kaya & Aycin, 2021) (Fig. 5).

3.5   Selection of Green Suppliers by Hybrid Fuzzy MCDM Approach (Çalık, 2021)

This study is presented by Ahmet Çalık in 2020. In this study, expert judgments are expressed using Pythagorean fuzzy numbers. Figure 6 depicts the structure of this study. The interval-valued Pythagorean fuzzy AHP method is used to calculate the weight of the criteria. The FTOPSIS method is used to calculate the distance between the positive and negative ideal solutions. A case study for the procurement of agricultural machinery is presented here (Çalık, 2021).

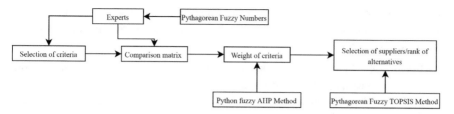

Fig. 6 Pythagorean fuzzy approach

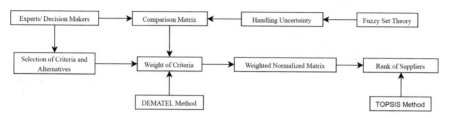

Fig. 7 Rough fuzzy approach

3.6 Selection of Sustainable Suppliers by Integrating Rough Fuzzy Approach (Chen et al., 2020)

The authors have proposed a novel framework in this work in order to identify smart sustainable SCMP (supply chain management practices) as selection of supplier. The given study is mainly divided into two parts, firstly, to compute the weight of both conventional and smart criteria and, secondly, to evaluate the ranking of suppliers. The workflow diagram is shown in Fig. 7.

DEMATEL (decision-making trial and evaluation laboratory) has been used to find the weight of the criteria as it provides effective interrelationship between criteria. TOPSIS (technique for order of preference by similarity to ideal solution) method is used to find the ranking of the suppliers because it consider both positive and negative criteria separately and provides fast solution. As a result, for smart supplier selection, hybrid MCDM, i.e., DEMATEL-TOPSIS method is utilized. The proposed strategy incorporates the strength of the fuzzy set in controlling internal uncertainty while utilizing the advantages of the rough set in manipulating external uncertainty (Chen et al., 2020).

4 Conclusion and Future Scope

The demand of the product of any industry depends upon various factors like quality, price, user-friendliness, shelf life, availability, etc. It depends on the features of raw material procured from the supplier. It is necessary to choose the right supplier for an efficient supply chain. According to the findings of this study, the conventional methods for supplier selection is different from those used in the era of Industry 4.0, in the following ways:

- *Uncertainty*: The present supplier selection methods can easily handle the condition of uncertainty by using the fuzzy set theory.
- *Real-time data*: Some models, such as MAS (multi-agent system), provide real-time data since this model provides a platform for manufacturers and suppliers to update their data, whereas conventional methods used existing data.
- *Dynamic*: As customer demand changes over time, supplier selection is becoming a critical issue in this dynamic situation. The presented approaches or models address this issue by allowing for real-time data analysis, which was not possible in conventional approaches.
- *Smart criteria*: The present study explains how the suppliers follow the smart supply chain, digitalization, automation, etc., by incorporating the criteria associated with Industry 4.0 like IoT, big data analytics,cloud computing, and digital collaboration.
- *Accuracy and time*: The conventional methods for supplier selection were less accurate and more time-consuming than the present models of Industry 4.0.

Many studies are performed on supplier selection in the era of Industry 4.0. There is a research gap for the selection of experts, weights assigning methods to the expert, which are not explained in the presented study. For future research, many more advanced or digital criteria can be incorporated, Machine learning and artificial intelligence can be used for precise results in big data analysis.

References

Abdel-Baset, M., Chang, V., Gamal, A., & Smarandache, F. (2019). An integrated neutrosophic ANP and VIKOR method for achieving sustainable supplier selection: A case study in importing field. *Computers in Industry, 106*, 94–110.

Büyüközkan, G., & Göçer, F. (2019). A novel approach integrating AHP and COPRAS under Pythagorean fuzzy sets for digital supply chain partner selection. *IEEE Transactions on Engineering Management.*

Çalık, A. (2021). A novel Pythagorean fuzzy AHP and fuzzy TOPSIS methodology for green supplier selection in the Industry 4.0 era. *Soft Computing, 25*(3), 2253–2265.

Chen, Z., Ming, X., Zhou, T., & Chang, Y. (2020). Sustainable supplier selection for smart supply chain considering internal and external uncertainty: An integrated rough-fuzzy approach. *Applied Soft Computing, 87*, 106004.

Ghadimi, P., Wang, C., Lim, M. K., & Heavey, C. (2019). Intelligent sustainable supplier selection using multi-agent technology: Theory and application for industry 4.0 supply chains. *Computers & Industrial Engineering, 127*, 588–600.

Gupta, H., & Barua, M. K. (2017). Supplier selection among SMEs on the basis of their green innovation ability using BWM and fuzzy TOPSIS. *Journal of Cleaner Production, 152*, 242–258.

Hasan, M. M., Jiang, D., Ullah, A. S., & Noor-E-Alam, M. (2020). Resilient supplier selection in logistics 4.0 with heterogeneous information. *Expert Systems with Applications, 139*, 112799.

Janvier-James, A. M. (2012). A new introduction to supply chains and supply chain management: Definitions and theories perspective. *International Business Research, 5*(1), 194–207.

Jayant, A., Gupta, P., Garg, S. K., & Khan, M. (2014). TOPSIS-AHP based approach for selection of reverse logistics service provider: A case study of mobile phone industry. *Procedia Engineering, 97*, 2147–2156.

Kaya, S. K., & Aycin, E. (2021). An integrated interval type 2 fuzzy AHP and COPRAS-G methodologies for supplier selection in the era of industry 4.0. *Neural Computing and Applications*, 1–21.

Li, Y. L., Ying, C. S., Chin, K. S., Yang, H. T., & Xu, J. (2018). Third-party reverse logistics provider selection approach based on hybrid-information MCDM and cumulative prospect theory. *Journal of Cleaner Production, 195*, 573–584.

Memari, A., Dargi, A., Jokar, M. R. A., Ahmad, R., & Rahim, A. R. A. (2019). Sustainable supplier selection: A multi-criteria intuitionistic fuzzy TOPSIS method. *Journal of Manufacturing Systems, 50*, 9–24.

Mousavi-Nasab, S. H., & Sotoudeh-Anvari, A. (2017). A comprehensive MCDM-based approach using TOPSIS, COPRAS and DEA as an auxiliary tool for material selection problems. *Materials & Design, 121*, 237–253.

Özbek, A., & Yildiz, A. (2020). Digital supplier selection for a garment business using interval type-2 fuzzy topsis. *Textile and Apparel, 30*(1), 61–72.

Prakash, C., & Barua, M. K. (2015). Integration of AHP-TOPSIS method for prioritizing the solutions of reverse logistics adoption to overcome its barriers under fuzzy environment. *Journal of Manufacturing Systems, 37*, 599–615.

Prakash, C., & Barua, M. K. (2016). A combined MCDM approach for evaluation and selection of third-party reverse logistics partner for Indian electronics industry. *Sustainable Production and Consumption, 7*, 66–78.

Sachdeva, N., Shrivastava, A. K., & Chauhan, A. (2019). Modeling supplier selection in the era of industry 4.0. *Benchmarking: An International Journal.*

Senthil, S., Srirangacharyulu, B., & Ramesh, A. (2014). A robust hybrid multi-criteria decision making methodology for contractor evaluation and selection in third-party reverse logistics. *Expert Systems with Applications, 41*(1), 50–58.

Shakourloo, A., Kazemi, A., & Javad, M. O. M. (2016). A new model for more effective supplier selection and remanufacturing process in a closed-loop supply chain. *Applied Mathematical Modelling, 40*(23–24), 9914–9931.

Singh, M., & Pant, M. (2021). A review of selected weighing methods in MCDM with a case study. *International Journal of Systems Assurance Engineering and Management, 12*, 126–144. https://doi.org/10.1007/s13198-020-01033-3

Sinha, A. K., & Anand, A. (2018). Development of sustainable supplier selection index for new product development using multi criteria decision making. *Journal of Cleaner Production, 197*, 1587–1596.

Stević, Ž., Pamučar, D., Puška, A., & Chatterjee, P. (2020). Sustainable supplier selection in healthcare industries using a new MCDM method: Measurement of alternatives and ranking according to compromise solution (MARCOS). *Computers & Industrial Engineering, 140*, 106231.

Xu, L. D., Xu, E. L., & Li, L. (2018). Industry 4.0: State of the art and future trends. *International Journal of Production Research, 56*(8), 2941–2962.

Xu, Z., Qin, J., Liu, J., & Martínez, L. (2019). Sustainable supplier selection based on AHPSort II in interval type-2 fuzzy environment. *Information Sciences, 483*, 273–293.

Yin, Y., Stecke, K. E., & Li, D. (2018). The evolution of production systems from industry 2.0 through industry 4.0. *International Journal of Production Research, 56*(1–2), 848–861.

Zekhnini, K., Cherrafi, A., Bouhaddou, I., Benghabrit, Y., & Garza-Reyes, J. A. (2020). *Supplier selection for smart supply chain: An adaptive fuzzy-neuro approach.*

A Comparative Approach for Sustainable Supply Chain Finance to Implement Industry 4.0 in Micro-, Small-, and Medium-Sized Enterprises

Pratik Maheshwari and Suchet Kamble

1 Introduction

Micro-, small-, and medium-sized enterprises (MSMEs) are the backbones of developing countries, and they play a crucial role in economic development (Kamble et al., 2020a). Since the 1990s, governments, practitioners, and researchers have been showing a renewed interest in re-examining the role of MSMEs due to economic globalization (Khanzode et al., 2021; MSMEs, 2021). Furthermore, economic globalization draws the government's attention to explore the challenges and opportunities associated with MSMEs for the country's overall development (Ghouri et al., 2021). The intensified competition in existing supply chains forces and influences the MSMEs to implement new technological advancements in their organization (Klapper, 2006; Sharma et al., 2021). The supply chain practitioners suggested that the Fourth Industrial Revolution (Industry 4.0) can mitigate the financial and operational challenges such as compliance, risk appetite, technological evolution, and lead time. However, Industry 4.0 technologies provide agility, resilience, cost benefits, flexibility, and business efficiency in the supply chain, whereas it creates innovation opportunities to ensure sustainable development of the country (Kamble et al., 2018b, 2020b & 2021b).

The new paradigm, "Sustainable supply chain finance," in the context of Industry 4.0 implementation, is a trending concept to achieve the prerequisite of the current competitive scenario (Paul et al., 2021; Tseng et al., 2018). Emerging cutting-edge technologies enable sustainable supply chain finance (SSCF) to mitigate the

P. Maheshwari (✉)
Indian Institute of Management, Jammu, India
e-mail: pratik@iimj.ac.in

S. Kamble
PV Vasantdada Patil College of Engineering, Mumbai, India

© The Author(s), under exclusive license to Springer Nature Switzerland AG 2023
S. S. Kamble et al. (eds.), *Digital Transformation and Industry 4.0 for Sustainable Supply Chain Performance*, EAI/Springer Innovations in Communication and Computing, https://doi.org/10.1007/978-3-031-19711-6_10

MSMEs' economic, environmental, and social responsibilities (Kamble et al., 2018b; Khanzode et al., 2021). Azadi et al. (2021) state that Industry 4.0 technologies are big data, blockchain, cloud computing, digital twin, artificial intelligence, and the Internet of things. Furthermore, SSCF can increase performance, safety, quality, and cost reduction (Jia et al., 2020). It also helps decrease supply chain expenditures and enhance its revenues (Jia et al., 2020). Sustainable finance (SF) has a crucial role in integrating the financial market actors, market policy, and related arrangements that can contribute to the inclusive development of the firm. The necessary SF instruments are contingent facilities, loan instruments, government aid/grants, (GA) internal finance, and bank loans (Gelsomino et al., 2016; Herath, 2015; Klapper, 2006).

The numerous capabilities and advantages of Industry 4.0 convinced the policymakers, government, researchers, and practitioners to develop and adopt those technologies in MSMEs (Abdel-Basset et al., 2020; Wang et al., 2021). In recent years, the topmost economies of the world, such as the USA, Germany, France, and China, entitled this Industry 4.0 implementation program named "Advanced Manufacturing Partnership," "High-Tech Strategy 2020," "La Nouvelle France Industrielle," "Made in China 2025," respectively (Azadi et al., 2021). A developing country such as India has designated the National Institute for Transforming India (NITI) Aayog to enable the policy framework and stabilize the public-private partnership model to initiate Industry 4.0 under the flagship of Samarth Udyog Bharat 4.0 (Khanzode et al., 2021; MSMEs, 2021).

Nevertheless, most MSMEs confront the adoption issues and barriers related to Industry 4.0 practices; hence, most of them are continuing with their traditional business and operational model due to fear of high implementation cost, lack of leadership, resource unavailability, lack of visibility, skills, and unawareness of schemes (Belhadi et al., 2021b; Sharma et al., 2021). Due to liabilities issues, most MSMEs are steadily ceding ground to larger firms (Klapper, 2006; Ma et al., 2020; Wuttke et al., 2019).

The selection of suitable SSCF resources is crucial for optimizing Industry 4.0 investment and implementation; these aspects have not drawn much attention in the research domain. Abdel-Basset et al. (2020) analyzed a novel decision-making model for SSCF and confirm that sustainability issues are still rare in the existing literature.

The combination of sustainability supply chain finance (SCF) with Industry 4.0 aims to upgrade the MSMEs in recent years (Ma et al., 2020; Tseng et al., 2018; Xu et al., 2018). The literature analysis presents that limited literature is available which addresses the solution strategy of the SSCF. Ding et al. (2015) and Paul et al. (2021) analyzed that data envelopment analysis (DEA) is the most powerful method for SCF in the literature. DEA is one of the efficient and rigorous nonparametric analytical methods that are highly flexible in simultaneously analyzing quantitative and qualitative measures (Dobos & Vörösmarty, 2019; Wu et al., 2021). In contrast, its application in SSCF is unexplored (Azadi et al., 2021). The objectives of this chapter are as follows:

1. To formulate the model for SSCF to implement Industry 4.0 in MSMEs
2. To explore SSCF strategies for Industry 4.0 under the zero inputs and ratio data for MSMEs
3. To incorporate loan amortization strategy under the various efficient decision-making units (DMUs)

This chapter has developed the SSCF model for Industry 4.0 implementation in the MSMEs concerning zero and ratio data. In comparison, our numerical illustrations incorporate a non-radial data envelopment model in the existence of ratio data and zero inputs along with its determined benchmarks on the Pareto efficiency frontier. Moreover, the proposed model can assist various government aid, internal financing, and external financing for the MSMEs to select the more efficient financing resources from the available set of resources. In addition, we provide a strategical road map to implement Industry 4.0.

The remainder of this chapter is as follows: The literature review is presented in Sect. 2. Section 3 explains the research method. The problem description and model formulation are provided in Sect. 4. The numerical illustration is shown in Sect. 5. The results and discussions are explained in Sect. 6, and Sect. 7 includes the managerial implication of the proposed model on MSMEs. Finally, the conclusions of this chapter are presented in Sect. 8.

2   Literature Review

We have performed bibliometric analysis to analyze SSCF in the context of Industry 4.0 implementation in MSMEs. We have followed the suggestion made by Krippendorff (2018) for sample generation. The relevant literature was extracted in four steps:

(a) *Definition of keywords*: The set of keywords belongs to one of four groups shown in Table 1. We have selected sustainable supply chain finance, Industry 4.0, and MSME-related keywords according to the theoretical foundations suggested by (Edwin Cheng et al., 2021; Kamble et al., 2018a; Khanzode et al., 2021; Xu et al., 2018). We do not refer to "sustainability" or "supply chain finance" or "Industry 4.0" or "MSMEs" as a single keyword because all the critical keywords themselves are a dynamic wide field of research. We have considered only papers containing at least one keyword of all the four groups in the title; abstract or list of given keywords were included. The literature search was performed in the Scopus database for these keyword combinations, resulting in 128 papers. Azadi et al. (2021) confirmed that integration of substantiable supply chain financing with Industry 4.0 is still in an infancy stage; our initial finding endorses a lack of literature available in this research domain.

(b) *Refinement*: This process only selected peer-reviewed English language journal papers. We further selected nine subject areas as relevant "environmental

Table 1 Sample keywords related to sustainability, SCF, Industry 4.0, and MSMEs

Sustainability (Edwin Cheng et al., 2021)	Supply chain finance (Xu et al., 2018)		Industry 4.0 (Kamble et al. 2018a, b)	MSMEs (Khanzode et al., 2021)
	Supply chain related	Finance related		
Sustainab*	Supply chain	Financing	Industrial internet of things	Small- and medium-sized enterprise
Eco*	Value chain	Trade credit	Cybersecurity	Entrepreneurship
Environment*	Logistics	Bank credit	Artificial intelligence	Fintech
Social	Inventory	Pay*	Big data	Business
Energy	Procurement	Capital constraint	Deep learning	SME-Manuf*
Pro-social	Purchasing	Factoring	Cloud computing	Service
Ethic*	Sourcing	Loan	3D printing	Organization

science," "economics, econometrics and finance," "mathematics," "engineering," "social sciences," "business, management, and accounting," energy," "computer science," "decision sciences," and leading to 65 research papers.

(c) *Inclusion and exclusion benchmark:* In this process, irrelevant papers were excluded by reading the title, keywords, and abstracts. After that, 32 papers remain in the sample after this process.

(d) *Snowball search:* All the articles were read completely and included 15 relevant papers addressing the financing strategy for the firms by backward snowball search, so 47 papers have remained in the final step.

2.1 Industry 4.0 and MSMEs

The concept of "Industry 4.0" refers to a new era of the digital industrial revolution to amalgamate organizational capabilities with ongoing sustainable technologies that provide an agile, flexible, resilient, and efficient environment for industries (Belhadi et al., 2021a; Kamble et al., 2020b). The pursuit of Industry 4.0 in industries mainly aims to deploy cyber-physical systems (CPS), smart manufacture, artificial intelligence, industrial Internet of things (IIoT), cognitive computing, machine learning, industrial robotics, big data, deep learning, reinforcement learning, deep neural networks, image processing, blockchain, and 3D printing with the fusion of advance industrial policies and their continuous improvement (Ghouri et al., 2021; Kamble et al., 2020b; Pundir et al., 2020; Sharma et al., 2021).

Figure 1 demonstrated the critical aspects of Industry 4.0. Kamble et al. (2020a) tabulated the Industry 4.0 technologies based on their purposes. The first purpose is to address smart data collections, analysis, and storage; this purpose can be fulfilled by prototyping, simulations, big data analytics, sensors, cloud computing, and the Internet of things (Luo & Choi, 2021). The second purpose addresses the shop floor

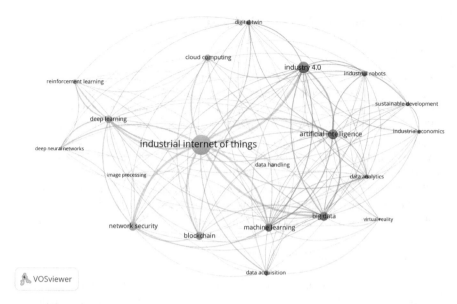

Fig. 1 Key aspects of Industry 4.0

technologies in which virtual reality, augmented reality, additive manufacturing, and robotics systems play a vital role. Finally, the third purpose explores integrating cyber-physical systems (CPS), digital twin, and cybersecurity technologies.

The concept of Industry 4.0 is evolving in MSMEs because it is the backbone of the developing countries because more than 80% of such industries are in rural areas, and especially in the Indian context, MSMEs employ more than 120 million people (CII, 2019). Khanzode et al. (2021) state that big firms usually outsource their work to MSMEs, and 95.98% of them are individual/family controlled/proprietary, so the management of MSMEs is centralized.

The MSMEs and small-medium enterprises (SMEs) are frequently used by researchers and practitioners in global and Indian contexts. The classification criteria of the concept are the fundamental difference between them (Khanzode et al., 2021). On the one hand, SMEs are measured by the number of employees and revenue generated by the firms, whereas at another hand, MSMEs are purely based on the plant and equipment's investment value (Khanzode et al., 2021). However, both types of industries face challenges and barriers to adopting the Industry 4.0 technologies.

Kamble et al. (2020b) periodized the enablers of Industry 4.0 for small and medium enterprises in the Indian context; their finding reveals that IIOT, big data analytics, and cloud computing have higher consideration, followed by additive manufacturing, augmented reality, and robotic systems. The MSMEs generally face organizational challenges to implementing digital technologies such as reluctant behaviors, immature leadership, financial constraints, low awareness, and lack of competencies (Ghouri et al., 2021). Ghouri et al. (2021) measure that the Industry 4.0 technologies enable the collaboration between different stockholders of MSMEs and supply chain partners to optimize the time, cost, and reliability.

2.1.1 Roadmap to Implement Industry 4.0 in MSMEs

• First Phase: Define.

Industry 4.0 technologies have a wide range of comprehensive digital technologies (See, Fig. 1). Due to various financial and operational constraints, implementation and deployment of each technology are not feasible for every MSME (Ada et al., 2021). According to the business problem, the manager should identify the project scope and vision (Yu et al., 2021). It requires the effective involvement of the company's external and internal stakeholders (Azadi et al., 2021). An appropriate selection strategy is mandatory for inter- and intraorganizational changes.

The manager can decide the project into three parts, i.e., short-term, medium-term, and long-term planning to achieve specific, measurable, achievable, realistic, and time-bound objectives (Ding et al., 2015; Dye & Yang, 2015; Wang et al., 2021).

• Second Phase: Measure.

The main aim of this phase is to identify the need for inter- and intraorganizational processes along with market conditions. This process typically includes data collection, collation, and analysis of digital technologies based on a diagnosis of the targeted problem with the help of statistical research; after this process, the value stream map identifies the cost-benefit of the whole process (Kamble et al., 2018a).

• Third Phase: Evaluate.

This process can incorporate the advantage of existing Industry 4.0 readiness assessment tools while assessing the current level of MSMEs readiness (Yu et al., 2021). This process aims to create a holistic view and quantitatively optimize the resources. Thus it helps to stabilize the strategy mapping for Industry 4.0 implementation (Herath, 2015).

• Fourth Phase: Optimize.

Nowadays, various optimization and simulation packages are available to optimize the project. The manager should optimize the requirement of Industry 4.0 technologies numerically and statistically. A digital twin can help in creating value to optimize resources (Kamble et al., 2018a).

• Fifth Phase: Develop.

The optimized detailed plan manager should incorporate a pilot project; based on the analysis, the essential project can implement at full scale. Wuttke et al. (2019) suggested that the implementation process should not disrupt the regular functionality of the firm.

• Sixth Phase: Validate.

The results from the pilot run should be analyzed effectively with stakeholders who should be under the baseline schedule and budgets (Jia et al., 2020; Tseng et al.,

2018). If results are not received significantly or unfavorable, the project team should justify the risk management strategy and replan the investments (Abdel-Basset et al., 2020).

- Seventh Phase: Implement.

Before implementing any project, the successful evaluation and validation of the pilot project report play an essential role in implementing a full-scale project. The manager should list the selected Industry 4.0 technologies and accommodate the project (Gornall & Strebulaev, 2018).

Recent studies categorized the Industry 4.0 implementation into three main clusters, i.e., smart factory, simulation and modelling, and digitalization and virtualization. The concept smart factory requires digital technologies such as embedded systems, Internet of services, product life cycle management, human-computer interaction, robotics, cyber-physical systems, RFID, additive manufacturing, automation, Internet of things, and modularization. The simulation and modelling involve mixed reality, augmented reality, and virtual reality. In contrast, digitalization and virtualization implementation required big data, mobile computing, cloud computing, and social media.

2.2 Supply Chain Financing and Sustainability

Figure 2 shows the foundational research on SCF by integrating the significant SCF definitions summarized by Wang et al. (2021). The first definition of SCF-I addresses SCF management in terms of optimized planning, controlling, cash flows, material flows, and managing facilities within the system (Gornall & Strebulaev, 2018; Wuttke et al., 2013). The second definition SCF-II presents the financing schemes to increase the monetary flow of capital (Gelsomino et al., 2016). Third, SCF definition merely refers to buyer-driven payable solutions (Klapper, 2006). This chapter adopts the second approach to analyze the SSCF in MSMEs.

Figure 3 demonstrated the most critical terms and linkage between various SSCF factors developed over the years. The analysis reveals that after 2018 supply chain finance, fintech, financial performance, and sustainability issues are the most influencing factors among the researchers and practitioners.

In the context of MSMEs, the win-win solution for lenders and borrowers emphasized that integration of overall supply chain stakeholders is essential instead of the particular echelon (Khanzode et al., 2021).

Wuttke et al. (2019) suggested that the benefits should be extended in terms of payments for progressive, reliable relationships for industrial development in MSMEs. Kamble et al. (2020b) analyzed the importance of the triple bottom line to measure the industry's financial, social, and environmental performance. Despite the capabilities and importance of SCF and sustainability, there is limited evidence on integrating both concepts (Belhadi et al., 2021b; Ma et al., 2020; Tseng et al., 2018).

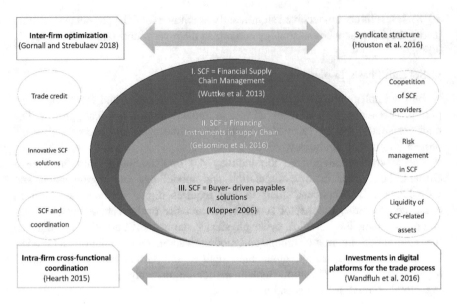

Fig. 2 Definitions of supply chain finance (Modified: Wang et al., 2021)

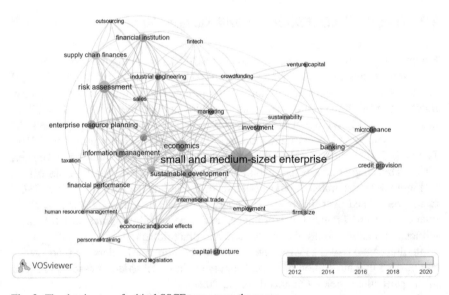

Fig. 3 The dominance of critical SSCF terms over the years

Belhadi et al. (2021a) and Kamble et al. (2020b) identified the barriers to adopting Industry 4.0 in MSMEs. Sharma et al. (2021) categorized adoption barriers in technology, organization, environment, and social and economic standard. The higher implementation cost, lack of financial support, and unawareness of government aid/grants are significant issues with MSME practitioners (Kamble et al. 2018a, b; MSMEs, 2021).

Table 2 Operations-finance interface for SCF

Major areas	Challenges	Prospects of improvements	References
Trade credit	Credit limit management	Reputation formation and bargaining power evolution	Wuttke et al. (2019)
SCF and coordination	Delays in reconciliation	Take advantage and interplay between multiple suppliers/retailers	Wang et al., 2021)
Innovative SCF solutions	Lack of financial flow	Systematic implementation and creates a catalyst for solutions	Gornall and Strebulaev (2018)
Coopetition of SCF providers	Lack of credentials	Better connects with banks for structure lending contracts	Houston et al. (2016)
Liquidity of SCF-related assets	Too much debt	Securitization and asset-backed securities	Khanzode et al. (2021)
Lending	Payment delays	Systematic capital investment	Xu et al. (2018)
Digital stock and distribution	Lack of facilities and skills	Upgrade to fintech solutions	Abdel-Basset et al. (2020)
Education of SC participation	Lack of training facilities	Credit guarantee trust fund for micro- and small enterprises (CGTMSE)	MSMEs (2021)
Digital payments	Manual processes	Interaction of exogenous and endogenous payments	Ma et al. (2020)
Data and risk analytics	Unsystematic information flow	Link the various stockholders on one platform and integrate both SC and non-SC settings	Kamble et al. (2020b)

Herath, 2015; Klapper, 2006). Notwithstanding significant challenges for MSMEs are the endowment, cash flow, risk perceptions, and lack of tailor-made solutions that result in lower credentials than large and public sector enterprises (Khanzode et al., 2021). The lack of real-time data on turnover assessment, inventory, asset and liability management, and mismatches in cash flows are the critical reasons for low credit scores for bank loans and government schemes (Jia et al., 2020; Ma et al., 2020). Therefore, in the absence of a credit rating, MSMEs struggle to get loans from the formal sectors. Table 2 shows the various operations-finance interface for SCF connecting different areas with their challenges and prospects of improvements concerning the existing literature.

This section has suggested some government aid/grants for technological upgradation with competitive (TUC) and quality certification in the context of Indian MSMEs for SSCF(Refer, Table 3).

MSMEs (2021) reported that repayment and the social issue had become significant challenges for lending institutions. Figure 4 demonstrated the indexed organizations of Indian MSMEs loans. Despite the available financial aid, the loan amount fluctuated between January 2020 and March 2021. These data indicate that social and organizational factors also influence the growth of MSMEs. Refer to the drastic change in period March 2020 to June 2020 due to COVID-19, where social

Table 3 Government schemes for TUC in Indian MSMEs (MSMEs, 2021)

Schemes	Description	Nature of assistance	Fund allocation in crores rupees (2020–2021)
ZED certification	The aim of this scheme is to implement zero-defect manufacturing and ensuring continuous improvement	Promote quality improvement in MSMEs	BE-51.75
ASPIRE	Create new jobs and reduce unemployment, along with facilitating an innovative business solution for unmet social needs	To ensure grassroots economic development at the district level	BE- 30 RE-15
NMCP	Capital subsidy for technology upgradation	Technological improvement and updation	BE- 503.28
EMDSI	Awareness and workshop program	Financial assistance 14 lakh to 1.00 crore rupees for procurement and installation of plant and machines	BE-50.09
QMS-QTT	Funding support for skill upgradation	Endeavors to sensitize MSMEs to adopt the latest QMS-QTT	BE-9
IPR	Building awareness on IPR	IPR tools for MSMEs	BE- 39.35
CLCSS	Credit linked capital subsidy scheme	Institutional finance for induction programs	BE-503.28
Digital MSMEs	To encourage the adoption of the IT and CT	To implement ERP, digital accounting, manufacturing and SC design, and regulatory compliances	BE- 85.705

Note: ZED, zero defect and zero effect; ASPIRE: A Scheme for Promotion of Innovation, Rural Industries, and Entrepreneurship; EMDSI, entrepreneurial and managerial development of SMEs through incubators; BE, budget estimated; RE, revised estimated; QMS & QTT, quality management standards and quality technology tools; IPR, intellectual property rights; CLCSS, credit linked capital subsidy scheme; IT & CT, information technology and communication technology

sustainability was the significant issue for the organizations. So apart from economic sustainability, social and organizational substantiality is also equally crucial for Industry 4.0 (Paul et al., 2021).

2.3 DEA and Decision-Making Units

Industry 4.0 financing is a complex process due to various constraints such as accommodation of multiplicity of inputs and outputs, compression between available best observations, lack of information about the prices, nonparametric variables, frontier complexity, and relative efficiency (Kamble et al., 2020a; Xu et al., 2018). To incorporate the practical constraints, it is necessary to measure the relative

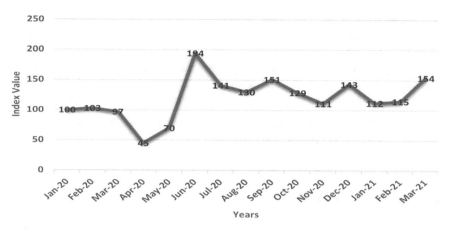

Fig. 4 Indexed originations of Indian MSMEs loans (₹ Lakh Crore) (MSMEs, 2021)

efficiencies of peer units (Ada et al., 2021). Therefore, A robust heterogeneous decision-making process is requisited to solve these challenges for MSMEs (de Oliveira Gobbo et al., 2021; Hatami-Marbini & Toloo, 2019; Kamble et al., 2021a). However, the DEA technique provides one possible solution approach to evaluate the relative efficiency of decision-making units (DMUs) under heterogeneous conditions (Andersen & Petersen, 1993; Wu et al., 2021). The DEA has proven to be a robust method for identifying the foremost practice frontier and aiding the acquisition limits (Ding et al., 2015). It does not require any a priori data about DMUs to solve multiple input and output problems (Wandfluh et al., 2016). Ding et al. (2015) state that it is a convenient technique for supplier selection, and their findings incorporate dual-role factors in a competitive environment.

In today's competitive and dynamic business environment, the significance of lender evaluation have become a critical issue. Dobos and Vörösmarty (2019) overcome this issues by using DEA for sustainable supply chain financing. Azadi et al. (2021) performed the bibliometric analysis; their cluster observation reveals that the DEA model can measure the sustainable financing resources for Industry 4.0 technologies, but it is in infancy stage.

2.4 Research Gap

The following key observations presents the knowledge gap based on the literature review.

1. Lack of literature on sustainable supply chain finance to mitigate competitive environment's challenges (Edwin Cheng et al., 2021; Jia et al., 2020).
2. Although there is a vast literature available on separate field such as sustainability, supply chain, finance, Industry 4.0, and MSMEs, the linkage is less uninvestigated and underdeveloped (Azadi et al., 2021; Lee, 2021).

3. Despite the vast and prominent applications of the DEA method, less research work is available on zero and ratio data (Azadi et al., 2021; de Oliveira Gobbo et al., 2021; Dobos & Vörösmarty, 2019; Lee, 2021; Wang et al., 2021).
4. Although there is availability of major government financial schemes for technological upgrading in MSMEs, strategic investment plans and their deployment are still lacking on the ground reality (Kamble et al., 2021b; MSMEs, 2021).

We have formulated the mathematical model and validated our findings with the existing studies to address the above research gap.

3   Research Method

Andersen and Petersen (1993) suggested the methodology to estimate the performance of DMUs and identified the ranks of efficient units with two inputs and one output strategy. However, their model is not applicable for superefficient DEA. Therefore, Seiford and Zhu (1999) improved the DEA method by performing an extreme efficient DEA referencing the endpoint of extreme efficient units and classifications, but their model could not evaluate the ratio data. Hollingsworth and Smith (2003) improved the existing DEA method by introducing ratios in DEA. Shafiee et al. (2014) proposed DEA based on a balanced scorecard approach to show the subunits' internal relationship. Lee (2021) proposed superefficiency and the slacks-based measures in the presence of nonpositive data, while Wu et al. (2021) incorporate the two-stage financing strategy for utility analysis.

Three types of DEA models are available in the existing literature based on radial, additive, and slack-based measure methods (Shafiee et al., 2014). We have followed the research method of Lee (2021) and Azadi et al. (2021). The first step is to create a production possibility set according to the actual observations of DMUs and then locate the DMUs according to the inputs. If the location point is not on the efficiency frontiers, then the DMUs are considered inefficient DMUs. To apply this method, first, we have to fix the direction of DMUs for improvement and evaluate the production possibility set.

4   Problem Description and Model Formulation

4.1   Problem Description

The resource allocation, budget, and lack of innovation practices are the significant challenges for adopting Industry 4.0 technologies and concepts for MSMEs (Khanzode et al., 2021). Literature analysis represents that up to now, fewer studies work on SSCF. However, the dynamic fields such as sustainability, supply chain financing, and Industry 4.0 are maturing over the years, but, still, the practical integration and deployment are untouched, while these integrations have tremendous potential for the growth of MSMEs (Azadi et al., 2021; Kamble et al., 2018b). This chapter mainly aims to identify the key Industry 4.0 technologies, sustainability issues with supply chain financing for MSMEs with zero inputs and ratio data, and estimation of loan amortization schedule. The literature analysis identifies these two main problems with the help of the DEA method.

4.2 Model Formulation

4.2.1 Notations

Particulars	Notations
DMUs	j
Input vector	i
Output vector	r
Intensity variable	λ_j
The ith input	$\bar{x}_i$
Objective function	ρ
The DMU under evaluation	DMU_o
Input of DMUo	x_{io}
Output of DMUo	y_{ro}
The rth output	$\bar{y}_r$
Linearize variable	t

The evaluation of different financial resources is the essential step to implement any decision-making process. In this chapter, we have considered three groups of financial resources, government grants and internal and external funding resources as a decision-making unit (DMUs).

We have incorporated the ratio data and zero input in our analysis. However, if we plot zero data on the efficiency frontier, it will be an infeasible solution. Therefore, DMU should be placed in a feasible region. Andersen and Petersen (1993) suggested that DMUs cannot be shift radially on the Farrell frontier by enhancing inputs. Due to this fact, superefficiency Russell and slacks-based measure models are infeasible for such cases (Azadi et al., 2021).

The model 1 incorporates zero inputs of DMUs and nonnegative constraints, whereas it can project the inefficient DMUs on the Pareto efficiency frontier. Eq. (6) of model 2 tries to decrease the nonproportionality of the inputs.

Model 1	Model 2
Objective function	Objective function
$\text{Min } \rho = \sum_{i=1}^{m_1} \bar{x}_i \ (i)$	$\text{Min } \rho_1 = \dfrac{\sum_{i=1}^{m_1} \bar{x}_i + \sum_{i=1}^{m_2} \bar{x}_i}{\sum_{i=1}^{m_1} x_{io} + \sum_{i=1}^{m_2} x_{io}} \ (1)$
Constraints	

(continued)

Model 1	Model 2
$\sum_{j=1,j\neq0}^{n} \lambda_j x_{ij} \leq \bar{x}_i, i = 1, \ldots\ldots m_1$, (ii)	Constraints
$\sum_{j=1,j\neq0}^{n} \lambda_j x_{ij} \leq \bar{x}_{io}, i = 1, \ldots\ldots m_2$, (iii)	$\sum_{j=1,j\neq0}^{n} \lambda_j x_{ij} \leq \bar{x}_i, i = 1, \ldots\ldots m_1$, (2)
$\sum_{j=1,j\neq0}^{n} \lambda_j y_{rj} \geq y_{ro}, r = 1, \ldots\ldots s_1$, (iv)	$\sum_{j=1,j\neq0}^{n} \lambda_j x_{ij} \leq \bar{x}_i, i = 1, \ldots\ldots m_2$, (3)
$\bar{x}_i \geq x_{io}, i = 1, \ldots\ldots m_2$, (v)	$\sum_{j=1,j\neq0}^{n} \lambda_j y_{rj} \geq y_{ro}, r = 1, \ldots\ldots s_1$, (4)
$\bar{x}_i \geq 0, i = 1, \ldots\ldots m_1$, (vi)	$\bar{x}_i \geq x_{io}, i = 1, \ldots\ldots m_2$, (5)
$\lambda_j \geq 0, j = 1, \ldots\ldots n, j \neq 0$, (vii)	$\bar{x}_i \geq x_{io}, i = 1, \ldots\ldots m_1$, (6)
	$\bar{x}_i \geq 0, i = 1, \ldots\ldots m_1$, (7)
	$\lambda_j \geq 0, j = 1, \ldots\ldots n, j \neq 0$, (8)

We have used theorem 1 to check the feasibility and significance of the proposed models 1 and 2 for a deeper understanding of the underlying concepts. Let us assume there is no internal dependability between the DMUs (Lee, 2021).

Theorem 1 Model 1 failed to present DMUs on the Farrell frontier while model 2 continuously provided a feasible solution.

Proof. If the ith input of the DMUs is equal to the ith input of DMU_o ($\bar{x}_i \geq x_{io}$, $i = 1, \ldots\ldots m_2$, $\bar{x}_i \geq x_{io}, i = 1, \ldots\ldots m_1$) where else this condition is followed by every production possibility set in that case, model 2 always has a feasible solution. Even though model 1 does not follow the requisite condition, so it is infeasible under these circumstances.

We used constraints 9 to concern the facilities for the nonnegative linear combination of DMUs in model 2.

$$\sum_{j=1,j\neq\alpha}^{n} \lambda^*_j x_{ij} \geq \lambda^*_2 x_{i2} \; i = 1, \ldots\ldots m_1 \& m_2 \tag{9}$$

Here the abbreviation α presenting the non-efficient DMUs. After the nonradial reduction in the inputs, we applied the additive model of Dobos and Vörösmarty (2019) to find the efficient and non-efficient DMUs by the following formula:

$$EFF.Set = \left\{ DMU_j | \rho^*_1 = 1 \right\} \; (\text{Model 3})$$

Seiford and Zhu (1999) proposed the model for a nonempty set, i.e.:

$$Nonempthy \; set = EFF.Set - \left\{ DMU_j | \rho^*_1 = 1 \right\} \; (\text{Model 4})$$

The main limitation of model1, 2, and 3 is that they do not consider the DMUs with zero inputs. To solve this contradiction, Lee (2021) proposed a model with at least one zero input. So, the updated model, i.e., model 5, expressed below:

$$Min.\rho_2 = \frac{\sum_{i=1}^{m_1} s_i^- + \sum_{i=1}^{m_2} s_i^+}{\sum_{i=1}^{m_1} x_{io} + \sum_{i=1}^{m_2} x_{io}}$$

Peixoto et al. (2020) suggested that benchmarking is a suitable tool for assessing and comparing the different models for DEA to apply that method; the efficiency score of each DMUs is required, which is presented by the following equation:

$$\nabla = \frac{\left(x_{input\_1} - s_1^{*-}\right) + \left(x_{input\_2} - s_2^{*+}\right)}{\left(x_{input\_1} + x_{input\_2}\right)}$$

The term ∇^* were used for optimal efficiency scores for n efficient and inefficient DMUs. Wu et al. (2021) proposed two-stage DEA models with fairness concern to a different real-world situation. The following equation expresses the integrated model:

$$Max\ \rho_3 = 1 + \left(\frac{1}{s}\right)\frac{\sum_{i=1}^{s} \overline{y_r}}{\sum_{i=1}^{s} y_{ro}}$$

So, the performance index of the particular DMUs will be given by:

$$\varnothing = \left(\frac{1}{2}\right)\left[\rho_1^* + \frac{1}{\rho_3^*}\right] \quad \text{(Model 6)}$$

We have used model 6 to get the most efficient DMUs, and then to evaluate the economic sustainability loan, amortization has been calculated and checked with the help of the following steps:

$$\mu = \frac{\sigma P}{n' \times \left[1 - \left(1 + \frac{\sigma}{n'}\right)^{-n't}\right]}$$

Where, P = principal amount, σ = rate of interest, t = time in terms of years, n' = monthly payment in a year.

Step 1: Declare data frame function considering loan amount, years, interest, annual percentage rate, and payment.
Step 2: Accept loan amount and duration.
Step 3: Calculate annuity factor for loan.
Step 4: Call data frames function.

5   Numerical Illustrations

In this chapter, we have developed the scenario on Indian MSMEs where the government funding agency is the ministry of MSMEs called GA in numerical illustrations. Let us assume the initial amount for Industry 4.0 implementation is estimated at USD 1,34,740.00. Thus, government aid and internal and external finance can provide the required investment in the MSMEs with different purposes and conditions. In general, financial practices, external financial sources, would like to charge the MSMEs with more interest rates. Nowadays, the inclusion of sustainable factors plays an essential role in selecting financial resources due to pressure from stockholders and government rules and regulations. As a result, the financial decision becomes a multi-criteria decision-making problem. We used loan amount, interest rate, and sharing profits as economic indicators to incorporate the current issues, with output targeting the environmental productivity and social productivity index as ratio data.

The dataset is shown in Table 4.

5.1   Solution Approach

The computation work is performed on R-version 4.1.0 with modified additive DEA, benchmarking, and multiplier DEA packages along with Microsoft Excel. To select the most profitable DMUs and deploy the sustainable implementing approach, first, we need to identify the inefficient DMUs. Models 1 to 4 failed to predict the inefficient units due to non-applicability issues with zero and ratio data, while model 6 provides the applicability to identify the most profitable DMUs with ranking facilities.

Table 4 The dataset on DMUs, inputs, and outputs

DMUs	Inputs			Outputs		
	Interest rate (%)	Amount ($) USD	Profit sharing (%)	Return period (month)	Environmental productivity index (EPI)	Social productivity index (SPI)
GA	2–7%[a]	53,896	0	300	19	12
Bank_1	10	26,948	0	150	10	9
Bank_2	9	20,211	0	130	12	7
Stakeholder_1	15	13,474	13	60	10	8
Stakeholder_2	12	13,474	12	84	9	7
Stakeholder_3	13	12,015	16	70	8	9
Individual	0	6737	59	55	13	10

[a] Depends upon the scheme mentioned in Table 3 (source: https://msme.gov.in/all-schemes)

Table 5 The results based on model 6

DMUs	ρ_1^*	ρ_3^*	$\frac{1}{\rho_3^*}$	$\varnothing$	Ranking
GA	1.037	0.885	0.8849	0.961	2
Bank_1	0.953	0.935	0.9345	0.944	3
Bank_2	0.793	0.959	0.9587	0.876	5
Stakeholder_1	0.528	0.980	0.9803	0.754	6
Stakeholder_2	0.844	0.950	0.9496	0.897	4
Stakeholder_3	0.191	0.975	0.9746	0.583	7
Individual	1	1	1	1	1

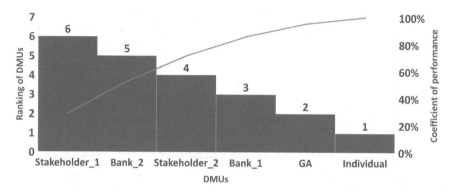

Fig. 5 Pareto chart for efficient DUMs

Table 5 identifies the ranking of the DMUs. Analysis reveals that stakeholder_3 is the most inefficient DMUs in this multi-criteria decision-making problem. Thus, the managers should avoid the loan with higher interest rates and higher profit-sharing stakeholders, whereas the government aids/grants get the second position in the ranking revels; that approach to government aid will always be beneficial for the managers. Figure 5 shows the ranking of the most efficient DMUs with a coefficient of performance. The curve shape of the Pareto line represents that a higher coefficient of performance value leads to a higher ranking of the DMUs.

6 Results and Discussions

The last column of Table 5 represents the ranking of the financial resources according to the DEA. On the given dataset, the ranking reveals that individual financing is the best source of the SSCF, although it is applicable only for the particular investment limit. Industry 4.0 implementation cannot proceed without external financial resources due to higher technological deployment costs. At the same time, government aid and the selection of sustainable financial resources make the objective more reliable and efficient. Figure 5 demonstrated that government aid, bank_1, stakeholder_2, and stakeholder_1 are the most efficient DMUs with the coefficient of performance 0.961, 0.944, 0.876, 0.754, and 0.897, respectively.

Table 6 represents the updation in the dataset based on model 5 in terms of slack additions. This process helps to locate the DMUs on Farrell efficiency frontiers.

The value should satisfy the optimal condition, i.e., $\left[\frac{1}{\rho_3^*}\right] \leq 1$ to compare the input-oriented efficiency from output-oriented efficiency, whereas the DMUs values near to score 1 is considered sustainable DMUs. Therefore, GA and bank_1 are the most sustainable DMUs in both input-output-oriented approaches. Furthermore, stakeholder_3 has a 0.583 value of the coefficient of performance which becomes the worst financial resource. Figure 6 shows the carrying value of loan amount over the years of different financial resources that reveal that managers should have to maintain the asset limit per year to repay the loan. Appendix shows the amortization schedule of the loan amount over the years.

7   Managerial Implications

The SSCF to implement Industry 4.0 technologies required higher various technical, social, and economic challenges. To sustain in the global market, managers should select efficient resources very carefully. As such, the SSCF ensures the management of financial pressures in terms of interest rates and flexibility. This chapter emphasizes government aid as an important financial resource for MSMEs. Furthermore, we have used zero as well as ratio data to analyze the efficiency of the available internal and external financial resources. Besides, evaluating the critical factors of sustainability managers can alleviate the pressure on government agencies and stakeholders.

In addition, the proposed model represents the amortization schedule of the loan amount over the years that helps the manager estimate the payback amount. The utilities of DMUs are expressed in terms of cooperative mode (refer models1 to 4) and noncooperative mode (refer models5 and 6). Models 5 and 6 represent real-life situations for DMUs.

The traditional models (models 1 to 4) are not applicable for ratio and zero data, whereas models 5 and 6 are applicable to ratio data and zero inputs. The contribution of this chapter is twofold, first to identify the available financial resources for MSMEs along with literature analysis on SSCF, which incorporate the Industry 4.0 and which we applied DEA with an input-output surplus for SSCF, and, second, the calculation of the nonradial DEA model for Industry 4.0 implementation and loan amount amortization.

Table 6 Slack variable addition based on model 5

DMUs	Score	Excess interest rate S-(1)	Excess amount (USD) S-(2)	Excess profit sharing (%) S-(3)	Shortage return period (month) S+(1)	Shortage environmental productivity index (EPI) S+(2)	Shortage social productivity index (SPI) S+(3)
GA	1.00	0.000	1.21266	0	0.00	0.000	0.00000
Bank_1	1.12	1.600	0	0	30.67	4.511	0.00000
Bank_2	1.00	0.000	0	0	0.00	0.000	0.00009
Stakeholder_1	1.10	3.319	0	0	29.18	0.000	0.00000
Stakeholder_2	1.00	0.000	0	0	0.00	0.000	0.00057
Individual	1.00	0.000	0	0	0.00	0.000	0.00000

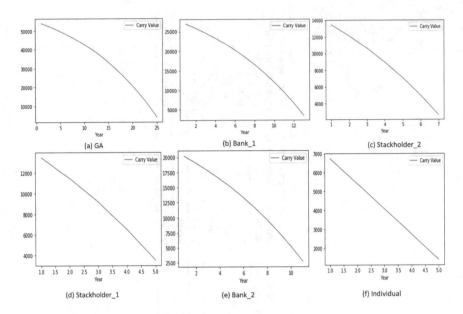

Fig. 6 Carrying value of loan amount over the years based on Table 4

8 Conclusions

Adopting Industry 4.0 in MSMEs will ensure the key benefits of efficiency, quality, cost-effectiveness, flexibility, and competitive advantages. While many MSMEs show a desire to deploy Industry 4.0, lack of knowledge and finance are the significant challenges for MSMEs, whereas sustainability issues make this decision more complicated. We have proposed the model to select the efficient DMUs with government aid, stakeholders, and banks. Furthermore, managers are reluctant to deploy digital technologies due to high implementation costs, while our analysis helps to provide a systematic process to select the most efficient source of finance with a sustainability index.

Our analysis shows that SSCF incorporating Industry 4.0 in MSMEs has ample opportunities for industrial growth and digitalization. We have provided the DEA model that can deal with zero and ratio data for practical implications. Moreover, we have considered the input-output surpluses to make the model more practical.

The main limitation of our model is that there is no internal exchange between the DMUs, but in a practical scenario, some internal dependability is always there between DMUs. More transparency is required to assess the sustainability issues in terms of the environmental productivity index (EPI) and social productivity index (SPI).

The proposed model can be extended with fuzzy data and determine the parameters for DMUs. The systematic investment plan for Industry 4.0 is not addressed and is apparent in the existing literature that could be incorporated in DEA more effectively. Furthermore, the clarity in the investment plan helps managers to allocate the fund in respective manners.

Appendix: Amortization Schedule of the Loan Amount Over the Years

Amortization Schedule: Bank_1

Date	Payment	Interest	Principal	Carry Value
2021-08-15				$26,948.00
2022-08-15	$3,793.70	$2,694.80	$1,098.90	$25,849.10
2023-08-15	$3,793.70	$2,584.91	$1,208.79	$24,640.31
2024-08-15	$3,793.70	$2,464.03	$1,329.67	$23,310.64
2025-08-15	$3,793.70	$2,331.06	$1,462.64	$21,848.01
2026-08-15	$3,793.70	$2,184.80	$1,608.90	$20,239.11
2027-08-15	$3,793.70	$2,023.91	$1,769.79	$18,469.32
2028-08-15	$3,793.70	$1,846.93	$1,946.77	$16,522.55
2029-08-15	$3,793.70	$1,652.26	$2,141.44	$14,381.11
2030-08-15	$3,793.70	$1,438.11	$2,355.59	$12,025.52
2031-08-15	$3,793.70	$1,202.55	$2,591.15	$9,434.37
2032-08-15	$3,793.70	$943.44	$2,850.26	$6,584.11
2033-08-15	$3,793.70	$658.41	$3,135.29	$3,448.82
2034-08-15	$3,793.70	$344.88	$3,448.82	$0.00

Total Interest Paid: $22,370.10

Amortization Schedule: Bank_2

Date	Payment	Interest	Principal	Carry Value
2021-08-15				$20,211.00
2022-08-15	$2,969.94	$1,818.99	$1,150.95	$19,060.05
2023-08-15	$2,969.94	$1,715.40	$1,254.53	$17,805.52
2024-08-15	$2,969.94	$1,602.50	$1,367.44	$16,438.07
2025-08-15	$2,969.94	$1,479.43	$1,490.51	$14,947.56
2026-08-15	$2,969.94	$1,345.28	$1,624.66	$13,322.90
2027-08-15	$2,969.94	$1,199.06	$1,770.88	$11,552.03
2028-08-15	$2,969.94	$1,039.68	$1,930.26	$9,621.77
2029-08-15	$2,969.94	$865.96	$2,103.98	$7,517.79
2030-08-15	$2,969.94	$676.60	$2,293.34	$5,224.45
2031-08-15	$2,969.94	$470.20	$2,499.74	$2,724.71
2032-08-15	$2,969.94	$245.22	$2,724.71	$0.00

Total Interest Paid: $12,458.33

Amortization Schedule: Individual

Date	Payment	Interest	Principal	Carry Value
2021-08-15				$6,737.00
2022-08-15	$1,388.09	$67.37	$1,320.72	$5,416.28
2023-08-15	$1,388.09	$54.16	$1,333.93	$4,082.35
2024-08-15	$1,388.09	$40.82	$1,347.27	$2,735.09
2021-08-15				$6,737.00
2022-08-15	$1,388.09	$67.37	$1,320.72	$5,416.28
2023-08-15	$1,388.09	$54.16	$1,333.93	$4,082.35
2024-08-15	$1,388.09	$40.82	$1,347.27	$2,735.09
2025-08-15	$1,388.09	$27.35	$1,360.74	$1,374.35
2026-08-15	$1,388.09	$13.74	$1,374.35	$0.00

Total Interest Paid: $203.45

Amortization Schedule: Stakeholder_1

Date	Payment	Interest	Principal	Carry Value
2021-08-15				$13,474.00
2022-08-15	$4,019.50	$2,021.10	$1,998.40	$11,475.60
2023-08-15	$4,019.50	$1,721.34	$2,298.16	$9,177.43
2024-08-15	$4,019.50	$1,376.61	$2,642.89	$6,534.54
2021-08-15				$13,474.00
2022-08-15	$4,019.50	$2,021.10	$1,998.40	$11,475.60
2023-08-15	$4,019.50	$1,721.34	$2,298.16	$9,177.43
2024-08-15	$4,019.50	$1,376.61	$2,642.89	$6,534.54
2025-08-15	$4,019.50	$980.18	$3,039.32	$3,495.22
2026-08-15	$4,019.50	$524.28	$3,495.22	-$0.00

Total Interest Paid: $6,623.52

Amortization Schedule: Stakeholder_2

Date	Payment	Interest	Principal	Carry Value
2021-08-15				$13,474.00
2022-08-15	$2,952.39	$1,616.88	$1,335.51	$12,138.49
2023-08-15	$2,952.39	$1,456.62	$1,495.77	$10,642.71
2024-08-15	$2,952.39	$1,277.13	$1,675.27	$8,967.45
2025-08-15	$2,952.39	$1,076.09	$1,876.30	$7,091.15
2026-08-15	$2,952.39	$850.94	$2,101.45	$4,989.69
2027-08-15	$2,952.39	$598.76	$2,353.63	$2,636.06
2028-08-15	$2,952.39	$316.33	$2,636.06	$0.00

Total Interest Paid: $7,192.75

Amortization Schedule: GA

Date	Payment	Interest	Principal	Carry Value
2021-08-15				$53,896.00
2022-08-15	$4,216.11	$3,233.76	$982.35	$52,913.65
2023-08-15	$4,216.11	$3,174.82	$1,041.29	$51,872.36
2024-08-15	$4,216.11	$3,112.34	$1,103.77	$50,768.60
2025-08-15	$4,216.11	$3,046.12	$1,169.99	$49,598.61
2026-08-15	$4,216.11	$2,975.92	$1,240.19	$48,358.42
2027-08-15	$4,216.11	$2,901.51	$1,314.60	$47,043.82
2028-08-15	$4,216.11	$2,822.63	$1,393.48	$45,650.34
2029-08-15	$4,216.11	$2,739.02	$1,477.09	$44,173.25
2030-08-15	$4,216.11	$2,650.40	$1,565.71	$42,607.54
2031-08-15	$4,216.11	$2,556.45	$1,659.65	$40,947.88
2032-08-15	$4,216.11	$2,456.87	$1,759.23	$39,188.65
2039-08-15	$4,216.11	$1,570.87	$2,645.24	$23,535.92
2040-08-15	$4,216.11	$1,412.16	$2,803.95	$20,731.97
2041-08-15	$4,216.11	$1,243.92	$2,972.19	$17,759.78
2042-08-15	$4,216.11	$1,065.59	$3,150.52	$14,609.26
2043-08-15	$4,216.11	$876.56	$3,339.55	$11,269.70
2044-08-15	$4,216.11	$676.18	$3,539.92	$7,729.78
2045-08-15	$4,216.11	$463.79	$3,752.32	$3,977.46
2046-08-15	$4,216.11	$238.65	$3,977.46	$0.00

Total Interest Paid: $51,506.68

References

Abdel-Basset, M., Mohamed, R., Sallam, K., & Elhoseny, M. (2020). A novel decision-making model for sustainable supply chain finance under uncertainty environment. *Journal of Cleaner Production, 269*, 122324. https://doi.org/10.1016/j.jclepro.2020.122324

Ada, N., Kazancoglu, Y., Sezer, M. D., Ede-Senturk, C., Ozer, I., & Ram, M. (2021). Analyzing barriers of circular food supply chains and proposing industry 4.0 solutions. *Sustainability, 13*(12), 6812. https://doi.org/10.3390/su13126812

Andersen, P., & Petersen, N. C. (1993). A procedure for ranking efficient units in data envelopment analysis. *Management Science, 39*(10), 1261–1264. https://doi.org/10.1287/mnsc.39.10.1261

Azadi, M., Moghaddas, Z., Farzipoor Saen, R., & Hussain, F. K. (2021). Financing manufacturers for investing in industry 4.0 technologies: Internal financing vs. external financing. International Journal of Production Research, 0(0), 1–17. https://doi.org/10.1080/00207543.2021.1912431.

Belhadi, A., Kamble, S. S., Mani, V., Venkatesh, V. G., & Shi, Y. (2021a). Behavioral mechanisms influencing sustainable supply chain governance decision-making from a dyadic buyer-supplier perspective. *International Journal of Production Economics, 236*, 108136.

Belhadi, A., Kamble, S., Gunasekaran, A., & Mani, V. (2021b). Analyzing the mediating role of organizational ambidexterity and digital business transformation on industry 4.0 capabilities and sustainable supply chain performance. *Supply Chain Management: An International Journal.*

CII. (2019). Making Indian MSMEs globally competitive. CII, the Mantosh Sondhi Centre; 23. Institutional Area, Lodi Road, New Delhi 110003, India. https://static.pib.gov.in/WriteReadData/userfiles/MSME.pdf

de Oliveira Gobbo, S. C., Mariano, E. B., & Gobbo, J. A., Jr. (2021). Combining social network and data envelopment analysis: A proposal for a selection employment contracts effectiveness index in healthcare network applications. *Omega, 103*, 102377. https://doi.org/10.1016/j.omega.2020.102377

Ding, J., Dong, W., Bi, G., & Liang, L. (2015). A decision model for supplier selection in the presence of dual-role factors. *Journal of the Operational Research Society, 66*(5), 737–746. https://doi.org/10.1057/jors.2014.53

Dobos, I., & Vörösmarty, G. (2019). Inventory-related costs in green supplier selection problems with data envelopment analysis (DEA). *International Journal of Production Economics, 209*, 374–380. https://doi.org/10.1016/j.ijpe.2018.03.022

Dye, C. Y., & Yang, C. T. (2015). Sustainable trade credit and replenishment decisions with credit-linked demand under carbon emission constraints. *European Journal of Operational Research, 244*(1), 187–200.

Edwin Cheng, T. C., Kamble, S. S., Belhadi, A., Ndubisi, N. O., Lai, K. H., & Kharat, M. G. (2021). Linkages between big data analytics, circular economy, sustainable supply chain flexibility, and sustainable performance in manufacturing firms. *International Journal of Production Research, 1–15*, 1. https://doi.org/10.1080/00207543.2021.1906971

Gelsomino, L. M., Mangiaracina, R., Perego, A., & Tumino, A. (2016). Supply chain finance: A literature review. *International Journal of Physical Distribution & Logistics Management., 46.*

Ghouri, A. M., Mani, V., Jiao, Z., Venkatesh, V. G., Shi, Y., & Kamble, S. S. (2021). An empirical study of real-time information-receiving using industry 4.0 technologies in downstream operations. *Technological Forecasting and Social Change, 165*, 120551. https://doi.org/10.1016/j.techfore.2020.120551

Gornall, W., & Strebulaev, I. A. (2018). Financing as a supply chain: The capital structure of banks and borrowers. *Journal of Financial Economics, 129*(3), 510–530. https://doi.org/10.1016/j.jfineco.2018.05.008

Hatami-Marbini, A., & Toloo, M. (2019). Data envelopment analysis models with ratio data: A revisit. *Computers & Industrial Engineering, 133*, 331–338. https://doi.org/10.1016/j.cie.2019.04.041

Herath, G. (2015). Supply-chain finance: The emergence of a new competitive landscape. *McKinsey on Payments, 8*(22), 10–16.

Hollingsworth, B., & Smith, P. (2003). Use of ratios in data envelopment analysis. *Applied Economics Letters, 10*(11), 733–735. https://doi.org/10.1080/1350485032000133381

Houston, J. F., Lin, C., & Zhu, Z. (2016). The financial implications of supply chain changes. *Management Science, 62*(9), 2520–2542. https://doi.org/10.1287/mnsc.2015.2159

Jia, F., Zhang, T., & Chen, L. (2020). Sustainable supply chain finance: Towards a research agenda. *Journal of Cleaner Production, 243*, 118680.

Kamble, S. S., Gunasekaran, A., & Gawankar, S. A. (2018a). Sustainable industry 4.0 framework: A systematic literature review identifying the current trends and future perspectives. *Process Safety and Environmental Protection, 117*, 408–425. https://doi.org/10.1016/j.psep.2018.05.009

Kamble, S. S., Gunasekaran, A., & Sharma, R. (2018b). Analysis of the driving and dependence power of barriers to adopt industry 4.0 in Indian manufacturing industry. *Computers in Industry, 101*, 107–119. https://doi.org/10.1016/j.compind.2018.06.004

Kamble, S. S., Gunasekaran, A., Ghadge, A., & Raut, R. (2020a). A performance measurement system for industry 4.0 enabled smart manufacturing system in SMMEs-A review and empirical investigation. *International Journal of Production Economics, 229*, 107853. https://doi.org/10.1016/j.ijpe.2020.107853

Kamble, S., Gunasekaran, A., & Dhone, N. C. (2020b). Industry 4.0 and lean manufacturing practices for sustainable organizational performance in Indian manufacturing companies. *International Journal of Production Research, 58*(5), 1319–1337. https://doi.org/10.1080/00207543.2019.1630772

Kamble, S. S., Belhadi, A., Gunasekaran, A., Ganapathy, L., & Verma, S. (2021a). A large multi-group decision-making technique for prioritizing the big data-driven circular economy practices in the automobile component manufacturing industry. *Technological Forecasting and Social Change, 165*, 120567.

Kamble, S. S., Gunasekaran, A., Subramanian, N., Ghadge, A., Belhadi, A., & Venkatesh, M. (2021b). Blockchain technology's impact on supply chain integration and sustainable supply chain performance: Evidence from the automotive industry. *Annals of Operations Research*, 1–26.

Khanzode, A. G., Sarma, P. R. S., Mangla, S. K., & Yuan, H. (2021). Modeling the industry 4.0 adoption for sustainable production in micro, small & medium enterprises. *Journal of Cleaner Production, 279*, 123489. https://doi.org/10.1016/j.jclepro.2020.123489

Klapper, L. (2006). The role of factoring for financing small and medium enterprises. *Journal of Banking & Finance, 30*(11), 3111–3130.

Krippendorff, K. (2018). *Content analysis: An introduction to its methodology* (4th ed.). Sage.

Lee, H. S. (2021). Slacks-based measures of efficiency and super-efficiency in presence of nonpositive data. *Omega, 103*, 102395. https://doi.org/10.1016/j.omega.2021.102395

Luo, S., & Choi, T. M. (2021). Operational research for technology-driven supply chains in the industry 4.0 era: Recent development and future studies. *Asia-Pacific Journal of Operational Research, 2040021*. https://doi.org/10.1142/S0217595920400217

Ma, H. L., Wang, Z. X., & Chan, F. T. (2020). How important are supply chain collaborative factors in supply chain finance? A view of financial service providers in China. *International Journal of Production Economics, 219*, 341–346.

MSMEs, G. O. I. (2021). Annual report 2020–21. Ministry of micro, small & medium enterprises. https://msme.gov.in/sites/default/files/MSME-ANNUAL-REPORT-ENGLISH%202020-21.pdf

Paul, A., Shukla, N., Paul, S. K., & Trianni, A. (2021). Sustainable supply chain management and multi-criteria decision-making methods: A systematic review. *Sustainability, 13*(13), 7104. https://doi.org/10.3390/su13137104

Peixoto, M. G. M., Musetti, M. A., & de Mendonça, M. C. A. (2020). Performance management in hospital organizations from the perspective of Principal Component Analysis and Data Envelopment Analysis: the case of Federal University Hospitals in Brazil. *Computers & Industrial Engineering, 150*, 106873. https://doi.org/10.1016/j.cie.2020.106873

Pundir, A. K., Ganapathy, L., Maheshwari, P., & Thakur, S. (2020). Interpretive structural modelling to assess the enablers of blockchain technology in supply chain. In *2020 11th IEEE annual information technology, electronics and Mobile communication conference (IEMCON)* (pp. 0223–0229). IEEE. https://doi.org/10.1109/IEMCON51383.2020.9284828.

Seiford, L. M., & Zhu, J. (1999). Infeasibility of super-efficiency data envelopment analysis models. *INFOR: Information Systems and Operational Research, 37*(2), 174–187. https://doi.org/10.1080/03155986.1999.11732379

Shafiee, M., Lotfi, F. H., & Saleh, H. (2014). Supply chain performance evaluation with data envelopment analysis and balanced scorecard approach. *Applied Mathematical Modelling, 38*(21–22), 5092–5112. https://doi.org/10.1016/j.apm.2014.03.023

Sharma, M., Kamble, S., Mani, V., Sehrawat, R., Belhadi, A., & Sharma, V. (2021). Industry 4.0 adoption for sustainability in multi-tier manufacturing supply chain in emerging economies. *Journal of Cleaner Production, 281*, 125013. https://doi.org/10.1016/j.jclepro.2020.125013

Tseng, M. L., Wu, K. J., Hu, J., & Wang, C. H. (2018). Decision-making model for sustainable supply chain finance under uncertainties. *International Journal of Production Economics, 205*, 30–36.

Wandfluh, M., Hofmann, E., & Schoensleben, P. (2016). Financing buyer–supplier dyads: An empirical analysis on financial collaboration in the supply chain. *International Journal of Logistics Research and Applications, 19*(3), 200–217.

Wang, J., Zhao, L., & Huchzermeier, A. (2021). Operations-finance interface in risk management: Research evolution and opportunities. *Production and Operations Management, 30*(2), 355–389. https://doi.org/10.1111/poms.13269

Wu, J., Xu, G., Zhu, Q., & Zhang, C. (2021). Two-stage DEA models with fairness concern: Modelling and computational aspects. *Omega, 102521*, 102521.

Wuttke, D. A., Blome, C., & Henke, M. (2013). Focusing the financial flow of supply chains: An empirical investigation of financial supply chain management. *International Journal of Production Economics, 145*(2), 773–789.

Wuttke, D. A., Rosenzweig, E. D., & Heese, H. S. (2019). An empirical analysis of supply chain finance adoption. *Journal of Operations Management, 65*(3), 242–261.

Xu, X., Chen, X., Jia, F., Brown, S., Gong, Y., & Xu, Y. (2018). Supply chain finance: A systematic literature review and bibliometric analysis. *International Journal of Production Economics, 204*, 160–173. https://doi.org/10.1016/j.ijpe.2018.08.003

Yu, Y., Zhang, J. Z., Cao, Y., & Kazancoglu, Y. (2021). Intelligent transformation of the manufacturing industry for industry 4.0: Seizing financial benefits from supply chain relationship capital through enterprise green management. *Technological Forecasting and Social Change, 172*, 120999. https://doi.org/10.1016/j.techfore.2021.120999

Artificial Intelligence and Data Science in Food Processing Industry

Mohit Malik, Vijay Kumar Gahlawat, Rahul S. Mor, Shekhar Agnihotri, Anupama Panghal, Kumar Rahul, and Neela Emanuel

1 Introduction

Even though the food industry is considered the most promising and competitive sector globally for generating employment opportunities, its significant reliance on human intervention for business operations faces fierce competition. The human workforce is essential to ensure smooth processing, packaging, and production of food items (Mor et al., 2022). Food processing industry requires a large amount of labor for manufacturing or servicing, and AI technologies help maximize productivity and reduce wastage. However, like other processing industries, ways to cost containment or expense reduction and striking a good balance between supply chains have been a major concern for food processing industries. Such quick decision-making requires using the sight and smelling senses and the ability to adapt to changing circumstances (Sharma et al., 2021). Although, technology is now employed to substitute human labor with replacing human work with smart

M. Malik · V. K. Gahlawat (✉) · K. Rahul · N. Emanuel
Department of Basic and Applied Sciences, National Institute of Food Technology
Entrepreneurship and Management, Sonepat, Haryana, India

R. S. Mor
Department of Food Engineering, National Institute of Food Technology Entrepreneurship and Management, Sonepat, Haryana, India

S. Agnihotri
Department of Agriculture and Environmental Sciences, National Institute of Food Technology Entrepreneurship and Management, Sonepat, Haryana, India

A. Panghal
Department of Food Business Management and Entrepreneurship Development, National Institute of Food Technology Entrepreneurship and Management (NIFTEM), Kundli, Haryana, India

© The Author(s), under exclusive license to Springer Nature Switzerland AG 2023
S. S. Kamble et al. (eds.), *Digital Transformation and Industry 4.0 for Sustainable Supply Chain Performance*, EAI/Springer Innovations in Communication and Computing, https://doi.org/10.1007/978-3-031-19711-6_11

devices. Food applications, robot deliveries, drones, and self-driving automobiles are all new ways to deliver information and food to consumers, and all of them rely on AI. It can help decrease the risk of diseases and detect contaminants in the food production chain to tackle the issues related to food quality and make the supply chain efficient (Keeble et al., 2020). It will also help provide "access to food" among individuals and ensure quality, which has been a goal of many intergovernmental policies throughout the globe. Developing new strategies and tools to improve food safety compliance and a constant effort toward compliance with international standards have been practiced by many food industries as it is directly related to country's economy (Kakani et al., 2020). Latest technologies such as AI and data science have played a vital role in achieving the desired results over the past few years (Misra et al., 2020). Therefore, it has become essential to investigate all aspects of the advanced food industry and intelligent agriculture. These technologies meet social needs and deliver the best quality products quickly. Modern technologies contribute toward scaling up production, thus increasing productivity and ensuring quality compliance which boosts the company's profit exponentially (Misra et al., 2020).

Autonomous or AI-based technologies are applied widely in almost every part of industry 4.0. It makes it easier to resolve the issues efficiently, automate the processes, and optimize production in food industries (Soltani-Fesaghandis & Pooya, 2018). The sector can ensure favorable crop monitoring, seed selection, temperature control, and watering with a computerized system to deliver excellent food products (Donepudi, 2014; Vadlamudi, 2018). The translational activities in the agri-food industry by implementing robotic farming, precision agriculture, and drone-based monitoring would foresee the importance of AI and machine learning technologies. The capabilities of AI are beyond that. It plays a vital role in food storage, processing, and delivery. Smart electronics such as drones and robots can be significant in reducing costs.

It is also helpful to deliver food items, finish the task in hazardous situations, and provide top-quality products (Bera, 2021; Castillo & Meliif, 1995; Tyagi et al., 2021). Two categories classify the roles of AI in the food industry. The first category deals with food quality control, such as the use of AI to improve food quality, use of mathematical modelling for food quality control, improvement in productivity of food products by using machine learning and AI in pesticide management. The second category deals with food hygiene, such as image detection and processing, fertilizer controls, food grading and inspection with image processing and AI, and robotics for food and warehouse safety (Mor et al., 2022).

It is essential for food businesses to cater to their customer's desires and needs to keep running smoothly in a saturated market. Choosing the ideal customer to target and finding out the needs of consumers in the food business is a challenge. Harnessing the power of data science can help satisfy customers and stand out in the competition. There are different ways it can leverage the opportunities in the food industry. With data science, one can analyze customer behavior and market trends. This data is beneficial to create daily procedures and schedules which are appealing and practical for potential shoppers.

AI is now a well-known concept among industries, product manufacturers, logistic service providers, suppliers, and advertisers (Pandian, 2019). By analyzing data, the businessman can make smart decisions according to customers' preferences. It demonstrates which things are in great demand in order to deliver the best products to customers. Data science can also help analyze customer traffic patterns to create the proper staffing schedules. Food establishments can be kept well-staffed to assist customers in rush hours instead of hiring many workers in off-peak hours. Households can easily order food quickly with several online food delivery partners such as Uber Eats, Zomato, and Swiggy. Food preparation, packaging, and delivery time must be managed effectively to provide on-time services.

Data analytics could be used for data governance and data collection to organize information and make it readily available to ensure the smooth performance of the business in food delivery. One can easily track and monitor orders to determine estimated arrival times for customers with data analytics. Food quality is the primary determining factor for 60% of customers before choosing a restaurant (Sophy, 2018). Serving best tasting and top-quality food is the primary factor in setting a business ahead of the competition. There is a thin line between having enough inventories to do customers without wasting and overbuying food. Proper big data governance can help sort data associated with readily available ingredients. It is essential to serve only fresh ingredients, considering the shelf life of such foods (DiSalvo, 2020). The key objectives of this chapter are to identify the role of AI, understand the importance of AI-based systems in solving food processing issues in the industry, and how AI and data science can revolutionize the food processing industry. This chapter is organized as follows. Section 1 begins with an introduction, Section 2 is the literature review, and Sect. 3 contains the methodology. Section 4 presents the findings and analysis, Sect. 5 includes the discussion, and Sect. 5 concludes this chapter.

2 Related Studies

The researchers (Lillford & Hermansson, 2021) observed that food science and technology is a significant knowledge base to permit advances in production to be sustainably converted to have enriched control over health with diet. New measurement science needs to be developed with better interdisciplinary collaborations. There is a global need to ensure food science and technology investment, but different approaches require application in local regions. Both private and public sector investments will boost training and education. It is vital to improve customers' awareness about modern technologies and their respective benefits in food supply systems and healthy diets. Sun et al. (2019) comprehensively review innovative drying technologies and their uses. Researchers introduced basic knowledge on expert systems, ANN, and fuzzy logic. ANN algorithms are used in smart city concepts (Basar et al., 2021). The authors summarized the application of AI in predicting, modelling, and optimizing mass and heat transfer, quality indicators, parameters of thermodynamic performance, and physiochemical elements of dried products. In addition, researchers also discussed the opportunities and challenges to find more research directions in this domain.

The food industry has faced many challenges regarding food safety and quality. There is an opportunity to overcome this issue in the existing technological environment where customers' choices and lifestyles are ever changing. Trust and transparency are the major drivers in the food chain for improvements and control in food integrity and economic growth (Dadi et al., 2021). The circular economy can drastically reduce wastage. Multiple hazards affect food commodities in the food chain cycle, and there is a high risk of contamination. These chemical and biological hazards are prevalent in food production, whether imposed fraudulently or brought accidentally. These hazards risk customers' faith in the food industry and affect their health. Along with the food chain and food safety monitoring solutions, a lot of data is generated with media, IoT, and other sensing devices. It is vital to use such data to help society (Nychas et al., 2021). Researchers discussed the safety and storage application of warehouses using robotics and Internet of Things (IoT) platforms, which may solve warehouse problems and minimize storage efficiency problems (Trab et al., 2018). Jiménez-Carvelo et al. (2019) published a comprehensive review of machine learning methods, representing food quality and analytical assessment. Several models for the food industry were developed. One was related to improving bread quality by enhancing sensory and nutritional components where different compositions were analyzed using principal component analysis (Qazi et al., 2022).

A vast amount of streaming data or "big data" is generated through the IoT, and it brings a lot of opportunities in food processes and agricultural monitoring. Social media also generates huge amounts of data, and it is also playing a vital role in the food industry. Misra et al. (2020) discuss the disruptive role of big data, IoT, and AI in defining the future of food industry solutions in supply chain modernization, food quality, and food safety. Along with the basics of big data, IoT, and AI, researchers also explored the role of big data and IoT in smart farm machines, greenhouse monitoring, and drone crop imaging in agriculture, social media in the food industry, supply chain modernization, food quality testing (sensor fusion and spectral methods), and food safety (with digital traceability using blockchain and gene sequencing) (Sivaganesan, 2021). Another model was proposed based on the AI cloud project, which designed an intelligent refrigerator that could send users an alarm if the food expired or was stored in the fridge for a long time (Nagaraju & Shubhamangala, 2020). Another method, named "an autonomous food wastage control warehouse," was created in 2020. The blockchain and numerous algorithms provided four primary benefits, including the warehouse giving priority to the food placed into the warehouse first and aiming to deliver it before it became stale food (Kumar et al., 2020).

A device known as the "electronic nose" was introduced in 2007. It worked similarly to the human nose, which was used to identify food quality. This was centered on sensors and was better than human sensory panels because it examined a large sample size and produced a more consistent and reliable result (Di Rosa et al., 2017; Schaller et al., 1998). Another technology known as the "electronic tongue" was developed, which had superior sensing and detection capabilities than people since it could recognize food odors and any dissolved substances (Tan & Xu, 2020).

Optimization and prediction and data mining techniques were commonly utilized in quality control, along with the ANN algorithm, an appropriate tool for improving food quality (Cao & Truong, 2016). Two models were used to compare the results and then utilize the best model to predict banana quality indices based on color attributes. The first was an ANN model for estimating banana quality, while the second was support vector regression (Sanaeifar et al., 2016). Toorajipour et al. (2021) discussed various AI techniques based on their application in supply chain management concerning applications. Monteiro and Barata (2021) presented a short overview of various AI algorithms frequently used in the agri-food supply chain in many areas such as production and monitoring. Goyache et al. (2001) discuss AI techniques in assessing food products. According to them, machine learning can find interpretable rules to classify samples despite the non-likeability of human processes or behavior, gather operative human knowledge from different examples, and determine the influence of each food on experts' decisions. Researchers define how to imitate the pattern of "bovine carcass classifiers," which will help further applications.

3 Methodology

Data is a supplementary strategy that investigators utilize for their research, and to put it another way, some researchers have already gathered and documented data for its long-term persistence. It can be found in government papers, books, journal articles, websites, and reports, among other places. Data for this chapter is collected from various databases such as Scopus, ScienceDirect, and IEEE. The primary objective is to ensure improved quality and food safety using data science and AI techniques. The authors have explored several relevant articles related to powerful AI techniques and machine learning algorithms to examine the role of AI in the food processing industry. Several areas in food processing are discussed in this chapter. Food security is the goal of the future food industry and agricultural sector. Using secondary data, this chapter is a detailed review of recent applications of AI in the food sector. AI could improve existing strategies and practices to achieve sustainability and positive results.

4 Applications of AI and Data Science in Food Processing Industry

This section explores various roles of AI in the food processing industry, such as sorting and packaging products, decision-making, new product launches, managing demand and supply, cleaning, and maintenance of equipment (Fig. 1). There are also many roles of AI in food handling where it handles the entire task of processing.

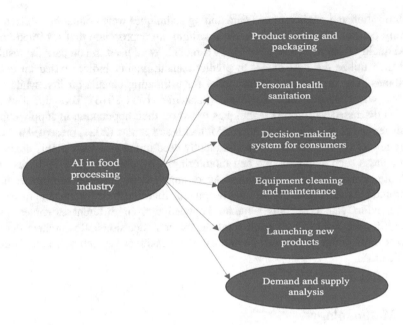

Fig. 1 Roles of AI in food processing. [Source:(Kumar et al., 2021)]

It can resolve the issues in real-time and shows promising results. Various AI algorithms and techniques are employed in the processing industry, such as the sorting process, cleaning and maintenance process, decision-making system to help consumers, predicting demands in supply chains, and helping in new product launches.

4.1 Sorting and Packaging of Products

Food ordering and packaging are tedious and time-consuming tasks in the food processing industry. Hence, AI-based solutions are used to reduce the risk of error and improve productivity in the industry. Deep learning approaches, which were first emerged in the 2010s and are implemented in various applications, represent the next stage in the growth of AI in sorting. These networks have been around for decades. It has become possible to apply them to actual problems because of a large increase in processing capacity in new graphics processing units and millions of widely available and tagged pictures (Tripathi et al., 2020; Wilts et al., 2021). During the COVID-19 pandemic with workforce level restrictions, AI and data science provided excellent sorting and packaging solutions. Several methodologies and tools are used in decision-making, such as laser-based systems, X-ray systems, high-quality cameras, and IR spectroscopy from harvest to packaging (Flemming & Balthasar, 2019). The flourishing food processing business uses AI and data science-based technology to utilize effective food resource management.

4.2 Decision-Making

It is observed that AI helps develop several food processing industries and choose innovative products for consumers (Dehghan-Dehnavi et al., 2020). Kellogg launched the "Bear Naked Custom" in 2018, which is ideal for consumers to make their granola adapted from 50 constituents. It records customer taste, flavors, and other details. That much data is enough for introducing a new product to the market, and AI has been helpful in customers' decision-making (Yost & Cheng, 2021). AI not just allows the industry to manage its services and goods but also assists consumers in making well-informed purchasing decisions. Customers could evaluate whether they want an item and later decide to use AI algorithms in surveys and advertising, unlike before (Liu et al., 2019).

4.3 Cleaning, Maintenance, and Sanitation

Another challenge for food manufacturers is keeping all food processing equipment clean. Proper maintenance and cleaning of processing equipment in food industries are essential. When food comes into contact, every piece of equipment and machine must be cleaned and disinfected thoroughly; if all processing is performed by AI-controlled robots and machines, removing humans from the process can help producers achieve a better standard of hygiene. AI-based tools can make it easy to handle (Wang et al., 2020). A lot of cameras and sensors are used for this task. The researchers explain its AI-based cleaning models in ultrasonic imaging and optical methods to gather the data (Schmidt & Piotter, 2020). Different countries such as the USA have issued guidelines regarding food processing sanitation. AI-based solutions have taken care of those guidelines. The Shanghai and Kankan health agencies collaborated to introduce AI-based solutions, and these solutions are designed to allow facial and object recognition for unknown numbers. The system can keep track of people who don't follow the guidelines (Rary et al., 2020). It can resolve the issues in real time and shows promising results.

4.4 New Launches

Launching a new product is not easy for a company, mainly associated with the food industry, relying entirely on what consumers want. Hence, a lot of decision-making systems collect customer data before launching a new product. The machine learning module processes the gathered information and makes decisions accordingly (Wardah et al., 2020). Coca-Cola has set up a self-service kiosk across the USA. This way, customers can make small changes in flavors to make plenty of customized drinks. The machine records such activities and deep learning, and machine learning algorithms perform the rest of the analysis. It is easy to launch new products with this data; Cherry sprite is a classic example (Kumar et al., 2021).

4.5   Management of Demand-Supply Chain

Supply chains have gotten increasingly difficult to manage in recent years. Physical flows are becoming longer and more interconnected as product portfolios become more sophisticated. Market volatility has increased the requirement for agility and adaptability, worsened by the COVID-19 epidemic (Olan et al., 2021). As a result of the rising focus on the environmental effects of supply chains, regional integration and flow optimization are becoming more common. Now industries and consumers focus more on supply-chain sustainability (Sharma & Patil, 2015). AI-based supply chain management platforms are predicted to be effective tools for assisting industries in addressing these issues. From procurement to distribution, an integrative edge framework can handle the possibilities and limits of all business areas. AI's ability to interpret massive amounts of data, identify linkages, offer transparency into processes, and assist in improved decision-making makes it a game changer (Cheraghalipour et al., 2018). AI has been used to keep track of every stage of the process. It controls everything, including the cost of inventory tracking. It also handles monitoring and forecasting of products from source to the market. Symphony Retail is an AI-powered facility to book billing, transportation, and inventory which also avoids ordering bulk products that may be wasted (Bhattacharyya et al., 2021).

4.6   Food Hygiene and Safety

Robots are widely accepted in the food sector because of their pure nature, and these are highly responsible for reducing food disorders. Strict hygiene guidelines have been issued by the "Food Safety Modernization Act (FSMA)" for overall supply chain operations. Spices, cereals, and other food items are kept in high vulnerability to contamination areas and cannot be refrigerated. This way, AI-based solutions can be helpful as their monitoring is easy and safe (Fedorova et al., 2020). Several other revolutionary approaches will soon come into prominence using AI in food safety. These can reduce the risk of food-borne illnesses. Food sellers are usually at the forefront of implementing new technologies that assist public health protection. However, AI and machine learning may not independently tackle all food safety issues, while they can be beneficial in resolving or helping with food safety issues. When deploying AI or ML, human error, security, and data inaccuracy are just a few of the fundamental issues retailers must consider (Hand, 2020). In order to keep the population safe, food safety-related information is quick and up-to-date which is extensive and the conclusion drawn are comprehensive. Hence, AI and data science play an important role in getting spot-on scientific information and depicting superior conclusions.

4.6.1   Next-Generation Sequencing and Electric Noses

The "electric noses (EN)" and "next-generation sequencing (NGS)" are some of the most promising technologies for food safety. Next-generation sequencing is the ideal alternative to DNA approaches in food safety. The onset of AI-based workflows and automated solutions has automated lab trials and data collection, and it has become more accurate and faster than ever. Harmful inclination can be found easily with NGS, and it can even avoid the spread of infection in public. On the other hand, electric noses are a substitute for fabrication. Some sensors can quickly identify different types of smells. These sensors can feel the scent and transfer the sensed information to the data center, where ML algorithms can access data (Yu et al., 2020). To improve the quality monitoring and testing of fruits and beverages, a range of e-nose devices are available. To optimize the required quality criteria, an appropriate sensor or a combination of sensors can be utilized to measure essential parameters for a particular application. Electronic tongue is highly beneficial for maintaining product quality standards (Mohamed et al., 2018).

4.7   Food Waste Management

According to the "Department of Agriculture, USA," around 30–40% of the food supply is wasted in the USA alone. The "Economic Research Service" by USDA estimated 31% of food waste to consumers at retail levels, that is, around 133 billion pounds would cost over $161 billion in 2010. That much waste has a significant impact on society (Kumar et al., 2021). According to McKinsey, AI could solve these issues related to food waste by 2030. Proper food management is required to deal with food waste, that is, proper resource utilization. It is the right time to replace conventional farming methods with more innovative approaches. ML algorithms can process the information collected from various sources and assist in making the right decisions (Filimonau et al., 2020). These algorithms help farmers to make better decisions. Here's how AI can reduce food waste by determining what microorganisms are healthy for vegetables and fruits to avoid using artificial fertilizers; AI can save considerable investment in ground surveys. It can examine and manage each process with computer vision; with the help of AI-powered food tracking, manufacturers can sell food before it is wasted. This way, food can reach more people, and farmers can increase productivity (Sharma & Abraham, 2020).

A lot of investment is needed in the food manufacturing and processing industry. AI technologies can identify many challenges in food manufacturing, unlike traditional systems. A lot of research is being done in AI and the food industry. Gayama is a classic example of a Swiss aggrotech firm as it has invested over $3.2 million for a project driven by AI. Researchers have designed "hyperspectral cameras" to scan even a tiny change in crop yields, water density, vermin, and nourishment. AI can also look for common threats and triggers signals to farmers to make decisions accordingly. It can suggest the proper measures that farmers can use available resources (SPD Group, 2021). It uses satellite data to analyze the earth's surface. Traditional farming practices have been no longer sustainable in several parts of the world, and smart farming could replace them soon. Smart farming can be made possible with AI to solve the existing problem of food waste (Filimonau et al., 2020). AI and ML are promising technologies to reduce waste in the food industry. The lab has developed ML-powered bots to pick food items from the plant to remove manual labor. These bots can analyze the maturity of fruit and detect fruits from various plants to handle them well (Kumar et al., 2021). Developing a robotic interaction platform with the multi-language capability to monitor and measure well-being, cloud-based drones used for real-time image processing and detection are also emerging (Kottilingam, 2020; Sungheetha & Sharma, 2020). Technologies like this are the future of farming and the food industry.

5 Conclusion

In the inception stage, AI and big data science modernize the food processing industry and have entered numerous food and beverage sectors. Automation is reshaping industries across the board, from procurement, waste treatment, and planned orders to forecasting the weather, food safety compliance, and new product development. AI plays a vital role in day-to-day life by improving food safety, hygiene, and waste management. It can transform the food processing sector to generate significant productivity for businesses. AI help eliminate much of the waste and inefficiency in our food systems, but we must design these tools for specific purposes. Thus, incorporating disruptive technologies is essential for the food processing industry. It was evident during COVID-19 when almost every sector faced difficulties operating and managing supply chains with a minimum workforce. The food industries need automation to tackle this problem and deal with situations such as COVID-19. In the future, a thorough examination of the different AI techniques used in the food industry will be a possible extension of this study.

Acknowledgments We acknowledge the National Institute of Food Technology Entrepreneurship and Management (NIFTEM), Kundli, Sonepat – 131028, India, for supporting this work.

Conflicts of Interest None.

Funding None.

References

Akey, S., & Sharma R. R. (2020). Real time monitoring and fire detection using internet of things and cloud based Drones. *Journal of Soft Computing Paradigm 2*(3), 168–174. https://doi.org/10.36548/jscp.2020.3.004

Basar, A., Smys, S., & Wang, H. (2021). 5G network simulation in smart citiesusing neural network algorithm. *Journal of Artificial Intelligence and Capsule Networks, 3*(1), 43–52. https://doi.org/10.36548/jaicn.2021.1.004

Bera, S. (2021). An application of operational analytics: For predicting sales revenue of restaurant. In S. K. Das, S. P. Das, N. Dey, & A.-E. Hassanien (Eds.), *Machine learning algorithms for industrial applications* (Vol. 907, pp. 209–235). https://doi.org/10.1007/978-3-030-50641-4_13

Bhattacharyya, S. S., Maitra, D., & Das, D. (2021). Study of adoption and absorption of emerging technologies for smart supplychain management: A dynamic capabilities perspective. *International Journal of Applied Logistics (IJAL), 11*(2), 14–54. https://doi.org/10.4018/ijal.2021070102

Cao, T. D., & Truong, H. L. (2016). Analyzing and conceptualizing monitoring and analytics as a service forgrain warehouses. In *Recent developments in intelligent information and database systems, studies in computational intelligence* (vol. 642, pp. 161–171). https://doi.org/10.1007/978-3-319-31277-4_14

Castillo, O., & Meliif, P. (1995). Automated quality control in the food industry combining artificial intelligence techniques with fractal theory. *WIT Transactions on Information and Communication Technologies.* https://doi.org/10.2495/AI950121

Cheraghalipour, A., Paydar, M. M., & Hajiaghaei-Keshteli, M. (2018). A bi-objective optimization for citrus closed-loop supply chain using Pareto-based algorithms. *Applied Soft Computing Journal, 69*, 33–59. https://doi.org/10.1016/j.asoc.2018.04.022

Dadi, K., Varoquaux, G., Houenou, J., Bzdok, D., Thirion, B., & Engemann, D. (2021). Population modeling with machine learning can enhance measures of mental health. *GigaScience, 10*(10), giab071. https://doi.org/10.1093/gigascience/giab071

Dehghan-Dehnavi, S., Fotuhi-Firuzabad, M., Moeini-Aghtaie, M., Dehghanian, P., & Wang, F. (2020). *Estimating participation abilities of industrial customers in demand response programs: A two-level decision-making tree analysis.* In 2020 IEEE/IAS 56th industrialand commercial powersystems technical conference (I & CPS), 1–8. https://doi.org/10.1109/ICPS48389.2020.9176817

DiRosa, A. R., Leone, F., Cheli, F., & Chiofalo, V. (2017). Fusion of electronic nose, electronic tongue and computer vision for animal source food authentication and quality assessment – A review. *Journal of Food Engineering, 210*, 62–75. https://doi.org/10.1016/j.jfoodeng.2017.04.024

DiSalvo, S. (2020). How data science plays a role in the food industry. *Master's in Data Science.* https://www.mastersindatascience.org/resources/food/. Accessed on 03 Feb 2022).

Donepudi, P. K. (2014). Technology growth in shipping industry: An overview. *American Journal of Tradeand Policy, 1*(3), 137–142. https://doi.org/10.18034/ajtp.v1i3.503

Fedorova, E., Darbasov, V., & Okhlopkov, M. (2020). The role of agricultural economists in study on problems related to regional food safety. *E3S Web of Conferences, 176*. https://doi.org/10.1051/e3sconf/202017605011

Filimonau, V., Todorova, E., Mzembe, A., Sauer, L., & Yankholmes, A. (2020). A comparativestudy of foodwastemanagement in full service restaurants of the United Kingdom and the Netherlands. *Journal of Cleaner Production, 258*. https://doi.org/10.1016/j.jclepro.2020.120775

Flemming, F., & Balthasar, D. (2019). An introduction to AI in sorting technologies. *Recycling Today.* https://www.recyclingtoday.com/article/an-introduction-to-ai-in-recycling-sorting-technologies/. Accessed on 03 Feb 2022.

Goyache, F., Bahamonde, A., Alonso, J., Lopez, S., del Coz, J. J., Quevedo, J. R., Ranilla, J., Luaces, O., Alvarez, I., Royo, L. J., & Diez, J. (2001). The usefulness of artificial intelligence techniques to assess subjective quality of products in the food industry. *Trends in Food Science & Technology, 12*(10), 370–381. https://doi.org/10.1016/S0924-2244(02)00010-9

Hand, A. (2020). Artificial intelligence advances food safety. In: profood world. https://www.profoodworld.com/processing-equipment/inspection/article/21195038/artificial-intelligence-advances-food-safety#:~:text=Landing%20AI%20is%20helping%20food,evaluate%20issues%20around%20food%20safety.&text=Machine%20vision%20has%20long%20found,working%2024%2F7%20without%20fatigue. Accessed 28 Jan 2022

Jiménez-Carvelo, A. M., González-Casado, A., Bagur-González, M. G., & Cuadros-Rodríguez, L. (2019). Alternative data mining/machine learning methods for the analytical evaluation of food quality and authenticity – A review. *Food Research International, 122*, 25–39. https://doi.org/10.1016/j.foodres.2019.03.063

Kakani, V., Nguyen, V. H., Kumar, B. P., Kim, H., & Pasupuleti, V. R. (2020). A critical review on computer vision and artificial intelligence in food industry. *Journal of Agriculture and Food Research, 2, 100033*. https://doi.org/10.1016/j.jafr.2020.100033

Keeble, M., Adams, J., Sacks, G., Vanderlee, L., White, C. M., Hammond, D., & Burgoine, T. (2020). Use of online food delivery services to order food prepared away-from-home and associated socio demographic characteristics: A cross-sectional, multi-country analysis. *International Journal of Environmental Research and Public Health, 17*(14), 1–17. https://doi.org/10.3390/ijerph17145190

Kottilingam. (2020). Emotional wellbeing assessment for elderly using multi-language Robot interface. *Journal of Information Technology and Digital World, 02*(01), 1–10. https://doi.org/10.36548/jitdw.2020.1.001

Kumar, I., Rawat, J., Mohd, N., & Husain, S. (2021). Opportunities of artificial intelligence and machine learning in the food industry. *Journal of Food Quality, 2021, 4535567*. https://doi.org/10.1155/2021/4535567

Kumar, V., Balasubramaniam, S., & Tharagesh, S. S. R. (2020). *An autonomous food wastage control warehouse: Distributed ledger and machine learning based approach.* In 2020 11th International Conference on Computing, Communication and Networking Technologies (ICCCNT). https://doi.org/10.1109/ICCCNT49239.2020.9225525.

Lillford, P., & Hermansson, A. M. (2021). Global missions and the critical needs of food science and technology. *Trends in Food Science and Technology, 111*, 800–811. https://doi.org/10.1016/j.tifs.2020.04.009

Liu, Y., Eckert, C., Yannou-Le Bris, G., & Petit, G. (2019). A fuzzy decision tool to evaluate the sustainable performance of suppliers in an agri food value chain. *Computers and Industrial Engineering, 127*, 196–212. https://doi.org/10.1016/j.cie.2018.12.022

Misra, N. N., Dixit, Y., Al-Mallahi, A., Bhullar, M. S., Upadhyay, R., & Martynenko, A. (2020). IoT, big data and artificial intelligence in agriculture and food industry. *IEEE Internet of Things Journal, 1*. https://doi.org/10.1109/JIOT.2020.2998584

Mohamed, R. R., Yaacob, R., Mohamed, M. A., Dir, T. M., & Rahim, F. A. (2018). Food freshness using electronic nose and its classification method: A review. *International Journal of Engineering and Technology, 7*, 49–53. https://doi.org/10.14419/IJET.V7I3.28.20964

Monteiro, J., & Barata, J. (2021). Artificial intelligence in extended agri-food supply chain: A short review based on bibliometric analysis. *Procedia Computer Science, 192*, 3020–3029. https://doi.org/10.1016/j.procs.2021.09.074

Mor, R. S., Kumar, D., Singh, A., & Neethu, K. (2022). Robotics and automation for agri-food 4.0: Innovation and challenges. In *Agri-food 4.0: Innovations, challenges and strategies.*

Nagaraju, T., & Shubhamangala, B. R. (2020). *Artificial intelligence powered smart refrigerator to arrest food wastage.* In 3rd international conference on innovative computing and com-munication (ICICC-2020). https://doi.org/10.2139/ssrn.3565256.

Nychas, G.-J., Sims, E., Tsakanikas, P., & Mohareb, F. (2021). Data science in the food industry. *Annual Review of Biomedical Data Science, 4*(1), 341–367. https://doi.org/10.1146/annurev-biodatasci-020221-123602

Olan, F., Liu, S., Suklan, J., Jayawickrama, U., & Arakpogun, E. (2021). The role of artificial intelligence networks in sustainable supply chain finance for food and drink industry. *International Journal of Production Research.* https://doi.org/10.1080/00207543.2021.1915510

Pandian, A. P. (2019). Artificial intelligence application in smart ware housing environment forautomated logistics. *Journal of Artificial Intelligence and Capsule Networks, 2019*(2), 63–72. https://doi.org/10.36548/jaicn.2019.2.002

Qazi, M. W., de Sousa, I. G., Nunes, M. C., & Raymundo, A. (2022). Improving the nutritional, structural, and sensory properties of gluten-free bread with different species of microalgae. *Food, 11*(3), 397. https://doi.org/10.3390/foods11030397

Rary, E., Anderson, S. M., Philbrick, B. D., Suresh, T., & Burton, J. (2020). Smart sanitation -biosensors as a public health tool in sanitation infrastructure. *International Journal of Environmental Research and Public Health, 17*(14), 1–14. https://doi.org/10.3390/ijerph17145146

Sanaeifar, A., Bakhshipour, A., & de La Guardia, M. (2016). Prediction of banana quality indices from color features using support vector regression. *Talanta, 148*, 54–61. https://doi.org/10.1016/j.talanta.2015.10.073

Schaller, E., Bosset, J. O., & Escher, F. (1998). "Electronic noses" and their application to food. *LWT – Food Science and Technology, 31*(4), 305–316. https://doi.org/10.1006/fstl.1998.0376

Schmidt, R., & Piotter, H. (2020). The hygienic/sanitary design of food and beverage processing equipment. In Demirci A., Feng H., Krishnamurthy K. (Eds.), *Food safety engineering* (pp. 267–332). Food Engineering Series. https://doi.org/10.1007/978-3-030-42660-6_12.

Sharma, S., & Patil, S. V. (2015). Key indicators of rice production and consumption, correlation between them and supply-demand prediction. *International Journal of Productivity and Performance Management, 64*(8), 1113–1137. https://doi.org/10.1108/IJPPM-06-2014-0088

Sharma, T. K., & Abraham, A. (2020). Artificial bee colony with enhanced food locations for solving mechanical engineering design problems. *Journal of Ambient Intelligence and Humanized Computing, 11*(1), 267–290. https://doi.org/10.1007/s12652-019-01265-7

Sharma, S., Gahlawat, V. K., Rahul, K., Mor, R. S., Malik, M. (2021) Sustainable innovations in the food industry through artificial intelligence and big data analytics. *Logistics, 5*(4), 66. https://doi.org/10.3390/logistics5040066

Sivaganesan, D. (2021). Performance estimation of sustainable smart farming with blockchain technology. *IRO Journal on Sustainable Wireless Systems, 3*(2), 97–106. https://doi.org/10.36548/jsws.2021.2.004

Soltani-Fesaghandis, G., & Pooya, A. (2018). Design of an artificial intelligence system for predicting success of new product development and selecting proper market-product strategy in the food industry. *International Food and Agribusiness Management Review, 21*(7), 847–864. https://doi.org/10.22434/IFAMR2017.0033

Sophy, J. (2018). Restaurant owners learn 60% of consumers will judge them on food quality. *Technology Trends.* https://smallbiztrends.com/2018/02/restaurant-experience-statistics.html. Accessed on 04 Feb 2022.

SPD Group. (2021). Machine learning and AI in food industry: Solutions and potential. https://spd.group/machine-learning/machine-learning-and-ai-in-food-industry/. Accessed on 03 Feb 2022.

Sun, Q., Zhang, M., & Mujumdar, A. S. (2019). Recent developments of artificial intelligence in drying of fresh food: A review. *Critical Reviews in Food Science and Nutrition, 59*(14), 2258–2275. https://doi.org/10.1080/10408398.2018.1446900

Tan, J., & Xu, J. (2020). Applications of electronic nose (e-nose) and electronic tongue (e-tongue) in food quality-related properties determination: A review. *Artificial Intelligence in Agriculture, 4*, 104–115. https://doi.org/10.1016/j.aiia.2020.06.003

Toorajipour, R., Sohrabpour, V., Nazarpour, A., Oghazi, P., & Fischl, M. (2021). Artificial intelligence in supply chain management: A systematic literature review. *Journal of Business Research, 122*, 502–517. https://doi.org/10.1016/j.jbusres.2020.09.009

Tripathi, S., Shukla, S., Attrey, S., Agrawal, A., & Bhadoria, V. (2020). Smart industrial packaging and sorting system. In Kapur P. K., Singh O., Khatri S. K., Verma A. K. (Eds.), Strategic system assurance and business analytics. Asset analytics (performance and safety management). https://doi.org/10.1007/978-981-15-3647-2_18.

Trab, S., Bajic, E., Zouinkhi, A., Abdelkrim, M. N., & Chekir, H. (2018). RFID IoT-enabled warehouse for safety management using product class-based storage and potential fields methods. *International Journal of Embedded Systems, 10*(1), 71–88. https://doi.org/10.1504/IJES.2018.089436

Tyagi, N., Khan, R., Chauhan, N., Singhal, A., & Ojha, J. (2021). E-rickshaws management for small scale farmers using big data-apache spark. *IOP Conference Series: Materials Science and Engineering, 1022*(1). https://doi.org/10.1088/1757-899X/1022/1/012023

Vadlamudi, S. (2018). Agri-food system and artificial intelligence: Reconsidering imperishability. *Asian Journal of Applied Science and Engineering, 7*, 33–42.

Wang, X., Puri, V. M., & Demirci, A. (2020). Equipment cleaning, sanitation, and maintenance. *Food Engineering Series, 333-353.* https://doi.org/10.1007/978-3-030-42660-6_13

Wardah, S., Djatna, T., & Yani, M. (2020). New product development in coconut-based agro-industry: Current research progress and challenges. *IOP Conference Series: Earth and Environmental Science, 472*(1). https://doi.org/10.1088/1755-1315/472/1/012053

Wilts, H., Garcia, B. R., Garlito, R. G., Gómez, L. S., & Prieto, E. G. (2021). Artificial intelligence in the sorting of municipal waste as an enabler of the circular economy. *Resources, 10*(4). https://doi.org/10.3390/resources10040028

Yost, E., & Cheng, Y. (2021). Customers' risk perception and dine-out motivation during a pandemic: Insight for the restaurant industry. *International Journal of Hospitality Management, 95*, 102889. https://doi.org/10.1016/j.ijhm.2021.102889

Yu, X., Lin, Y., & Wu, H. (2020). Targeted next-generation sequencing identifies separate causes of hearing loss in one deaf family and variable clinical manifestations for the p.R161C mutation in SOX10. *Neural Plasticity, 2020,* https://doi.org/10.1155/2020/8860837.

Industry 4.0-Based Agritech Adoption in Farmer Producer Organization: Case Study Approach

C. Ganeshkumar, A. Sivakumar, and B. Venugopal

1 Introduction

Farmer producer organizations (FPOs) are new organizational forms promoted by the Government of India through NABARD and SFABC. These FPOs have been encouraged to enhance the bargaining power of small and marginal farmers in obtaining their inputs for agriculture and in selling their outputs in the market. Small and marginal farmers had otherwise the option of cooperatives, which were under a different statute and were constrained due to several issues over years (Siddhartha et al., 2019). One of the most successful cooperative brands is Anand Milk Union Limited (AMUL). AMUL represents the brand of a milk marketing federation in Gujarat that has gone national. Replicating the AMUL model, there have been successful regional brands established in several states across India. These dairy cooperatives have been not successful in helping the small and marginal dairy farmers. The liberalization of the dairy sector had resulted in development of large dairy farms and the increased role of the private sector in dairy development (Pachayappan et al., 2020). Several Indian and international dairy brands have expanded in the recent years. FPOs provide an opportunity for the small and marginal dairy farmers who own 2–3 milch animals to use a new organizational form to enhance their bargaining power in obtaining feed, other inputs, and services. This power could be utilized for obtaining a higher price for milk too (David, 2020).

C. Ganeshkumar (✉) · B. Venugopal
Indian Institute of Plantation Management, Bangalore, Karnataka, India

A. Sivakumar
VIT Business School, VIT Vellore, Vellore, Tamil Nadu, India
e-mail: sivakumar.a@vit.ac.in

© The Author(s), under exclusive license to Springer Nature Switzerland AG 2023
S. S. Kamble et al. (eds.), *Digital Transformation and Industry 4.0 for Sustainable Supply Chain Performance*, EAI/Springer Innovations in Communication and Computing, https://doi.org/10.1007/978-3-031-19711-6_12

Agritech using information technology applications in agriculture in recent times has grown as a specialized technology. The use of these digital technologies in different value chain stages of agriculture can result in several benefits. Agritech companies joining hands with FPOs can lead to greater synergy in several dimensions like better data collection, monitoring, and analysis resulting in value both for the agritech companies and the FPOs. Specifically digital technologies in dairying can help in management across multiple aspects of dairy management like buying cattle, managing the environment for cattle growing, monitoring cattle health, and optimizing feed and other inputs to increase the productivity of the animals (Ganeshkumar et al., 2022). Research is available in the case of FPOs individually and about agritech companies individually both in India and the international contexts. However, there is hardly any research on the interaction between FPOs and agritech companies. Since both these are new developments in Indian agriculture, a study of this question on whether agritech can help support FPOs in their development is an aspect that is worth an exploration (Alagh, 2019). Through this research, various dimensions of agritech and FPO that have an influence on the synergy can be explored. This could lead to policy implication for the governments and the implementing institutions in developing these two sectors. Moreover, the study can help in providing specific managerial implications for both the agritech companies and the FPOs (David et al., 2022).

2 Farmer Producer Organization

Producer organizations have been existent in different forms across the world. Specifically, producer cooperatives became popular with the cooperative movement. The cooperative movement became popular in the agricultural sector; as in most countries, individual farmers dominate this sector. Depending on the country, the farm area holding of the farmers differed (Kanitkar, 2016). In India, with the predominance of small and marginal farmers, agricultural cooperatives came into prominence during the 1960s. Until the late twenty-first century, agricultural cooperatives dominated the agricultural scene, in aggregating small and marginal farmers, across different subsectors of agriculture like crop and animal farming. In the beginning of the twenty-first century, efforts at looking to different other organizational forms for consolidating the small and marginal farmers took place (Mahajan, 2014). This led to the development of FPOs in India. FPOs are businesses recognized as companies under MCA. Different types of FPOs exist. For example, crop-specific or agricultural activity-specific FPOs like turmeric growing farmers starting FPO or dairy farmers forming a FPO are common. In addition, the nature of the organizational form of the FPO depends on the promoting or resource institution (Govil et al., 2020).

3   Agritech Companies

Agritech companies use digital technologies to increase agricultural yield, quality, efficiency, and profitability. Accurate weather forecasts, automated irrigation and harnessing data collected through smart technologies through analytics, and machine learning help farmers to make better decisions in the agriculture (Prasad, 2017). The use of drones and satellites to scan crops in the fields in order to measure and monitor its performance is another use of agritech. They also use digital technologies in disease protection, pollination, planting seeds, and food security (Sowmya & Raju, 2017). Another technology is Internet of things (IoT) that uses software to help with information on weather, humidity, and current condition of soil. In addition, blockchain technology help trace origins of produce and track the journey from seed to final produce. Moreover, several companies are using AI and data analytics to analyze the vast information already available about the land and crop from the farmer (Trebbin, 2016). In the context of dairying, dairy digital tech companies facilitate feed bunk management, cow behavior monitoring, cow health, milk quality, and manure treatment. In addition, certain companies are focused on milk and milk product including value-added product delivery to the final consumers. Thus, the entire value chain of dairying is impacted through dairy technology companies (Siddhartha et al., 2021).

In India, dairying is a major business, which developed with the encouragement of the government. It mainly consists of the small and marginal farmers growing a small number of milch animals (less than five). Considering this situation this study is done with following research questions and objectives with a focus on a dairy tech company. The research question is to see how FPOs would obtain greater benefit for small and marginal farmers with the use of agritech. The research objectives are to analyze the issues that small and marginal farmers face in different stages of agriculture, to understand FPOs contribution in achieving greater negotiating power for small/marginal farmers and to decipher the role of agritech in facilitating FPOs (Ganeshkumar et al., 2021).

4   Literature Review

The review of studies on the topic can be categorized into studies on FPOs, studies on agritech companies, and studies that talk of the interaction of FPOs with agritech companies.

4.1   Studies on FPOs

Raju et al. (2017) studied FPOs in Andhra Pradesh. This was a study to assess the scope of developing FPOs in the state. Kumar et al. (2019a, b) studied how FPOs affected organic chili production. Similarly, Gokul et al. (2019) analyzed the roles of various actors in a FPO-based millet value chain. Restricting themselves to the processing subsector, Verma et al. (2017) studied the role of FPOs in promoting processing. Taking a different approach to FPO studies, Suriyapriya et al. (2019) looked at a particular service, namely, mobile advisory service, and studied the perception of FPO members. An interesting dimension of FPOs is their ability to act like business entities. Deka et al. (2020) studied vegetables from FPOs in West Bengal to understand this aspect. From an agricultural extension point of view, FPOs help in facilitating agricultural extension; this point of view was studied by Krishna (2018). Padmaja et al. (2019) in their study did a macroanalysis of FPOs in India. They studied the trends and patterns of the development of FPOs and offered suggestions for future development. Taking a district-level approach, Kathiravan et al. (2020) analyzed the communication and usefulness of FPOs among farmer members. Jose and Meena (2019) studied the profile of a Kerala-based dairy producer company's members to understand the various demographic factors that may influence decision-making (Ganeshkuma et al., 2020).

Verma et al. (2019) used Bihar state as a unit of analysis in finding out if FPOs have been beneficial to farmers. Ramappa and Yashashwini (2018) in contrast studied the historical development of FPOs in India and analyzed the challenges and opportunities. An important aspect of FPO analysis is the understanding of financial sustainability. Kakati and Roy (2019) in their study analyzed this aspect of sustainability using FPOs in Northeast India. Vijayakumar (2020) using a crop-based approach studied the effect of a coconut FPO in its impact on rural Karnataka. Kumar et al. (2019a, b) studied how the transformation of a farmer from a grower to a seller was facilitated through FPOs. Vishnu and Gupta (2017) analyzed the innovative dimension of FPOs in a local system through their research.

4.2 Agritech Adoption

Digital technologies in agriculture have been ubiquitous with the digitalization of various dimensions of the farming business. The developed countries have seen the use of these technologies quite early compared to the developing countries like India. Giaffreda (2019) conducted a broad and interesting study on the various technologies available and why farm businesses and farmers adopt these technologies. It looks at incentives that have spurred adoption of agritech. Payne-Gifford et al. (2020) in their cross-country analytical paper explored what attitudes triggered adoption of agritech. Amato et al. (2021) in their paper looked at a specific technology, namely, AI in the agritech domain (Ganeshkumar et al., 2019).

Agriculture is subject to the vagaries of the natural environment, and therefore technology can help predict changes in the nature to suit better farming. In their paper, Simelton and McCampbell (2021) explored the role of climate prediction through technology in facilitating climate resilient farming. An important influencer of technology adoption in any sector and specifically agriculture are policies and funding. Serrano et al. (2020) in their research attempted the analysis of these aspects on climate change mitigation. Adoption of agritech differs across countries. In an interesting paper, Knierim et al. (2018) studied them across multiple countries in a cross-country study. To be more specific adoption changes across types of farmers targeted for these technologies. Kendall et al. (2021) did a China-based family small farm study on agritech adoption (Victer Paul et al., 2020).

Operationally drones are being used extensively in developed countries for various agricultural operations. This was the focus of the research of Spanaki et al. (2021), and they inked it to better food security. A major benefit of digital technology introduction in agricultural is the ease of information exchange among the various actors in farming. Dolfsma et al. (2021) studied agricultural supply chain for their effect in information exchange with the introduction of agritech. Design of agritech requires working with the farmer community that would be using the technology. This would help in better technology adoption. In their research, based on app development for agricultural use, Kenny et al. (2021) suggested a design thinking-based method for technology development. The emphasis globally is on developing sustainable agricultural systems. In tune with this, Finger et al. (2020) researched on the contribution of agritech to agricultural system sustainability. The above review of recent literature on FPOs and agritech shows that there exists a gap in studying adoption of agritech by FPOs. There are several dimensions of this interaction that needs to be studied so that the question of synergy between the two can be established (Attuario et al., 2019).

5   Research Methodology

Case research initially started as a qualitative research method and later got integrated in quantitative and mixed method research studies. Case research typically follows one of the four schools of research of the pioneers in the field, namely, Yin, Eisenhardt, Stake, and Merriam. Yin and Eisenhardt are typically positivist in approach. As per Yin, 2018, when the phenomenon is contemporary and needs to be studied in depth but is difficult to be distinguished from the context, then case research is the most suitable form of research. In the current study, the synergy due to the interaction between the FPO and the agritech firm is a contemporary phenomenon, and since the context and the phenomenon are difficult to distinguish, case research was considered appropriate. In addition, appropriate existing theories would be used to provide the possible explanations. Data for the case came primarily from secondary sources and short interviews with resource institutions and FPOs (Shah, 2016).

Following Yin's suggestion, this study uses explanation building as the method for analysis in the current research. In addition, this research uses the single holistic case study studying the agritech firm Dvara E-Dairy solutions. As the phenomenon studied is new, the single case study would also be of revelatory nature. Causal and rival explanations would be used to increase the validity of the study. The case study approach used matches with the typical requirements of Kathleen Eisenhardt (Eisenhardt, 2021) especially the lack of empirical evidence in studying the phenomenon. Following Robert Stakes' (Stake, 1988), interpretivist/constructivist approach, this case research can be termed as instrumental case where the case is a general one to understand the context and the phenomenon. In addition, the research question is more of the issue question that is the exploration of the nature and extent of adoption of agritech among FPO members. Interviews, observations, and documents have been used for data capturing in the qualitative as well as quantitative form.

As per Sharan Merriam's (Merriam & Grenier, 2019), case approach should be particularistic and descriptive. Moreover, it would be an intrinsic and instrumental case. FPOs and agritech are new phenomena in the Indian context. In such a situation, when the phenomenon is broad and complex and extant research is insufficient for posing causal questions, case research is suggested. Moreover, holistic and in-depth investigation is typically possible only in case research to understand the nuances of interaction between the two phenomena (Panpatte & Ganeshkumar, 2021). In addition, case research allows for the role of the researcher to be a detached observer. If an in-depth, multifaceted understanding of a complex issue in its real-life context is experienced, case research is always the best research method for pursuing research. In view of these reasons, this research is planned as a single case research on an agritech company called Dvara E-Dairy. The study involved in-depth discussion with several senior officials of Dvara E-Dairy and in-depth interviews with dairy FPOs which had both adopted and not adopted the new age digital dairy technology for feed and health management (Paul et al., 2019).

6   Results and Discussion

Livestock forms a major part of Indian agriculture contributing to more than 30% of the gross value of output in agriculture and allied sector. In addition, small and marginal farmers (area less than 0.01 hectare landholding) constituting more than 20% of the total farmers depended on livestock as a major income. The importance of dairying within livestock as a sector can be gauged from its output value exceeding combined output value from cereals, pulses, oilseeds, and sugarcane in 2018–2019. In 1991, the milk sector was liberalized, but the complete removal of all restrictions came in 2003 resulting in a huge capacity addition to milk processing. Dairy cooperatives lead the pasteurized liquid milk market. Milk production doubled in 15 years between 2003 and 2018. Similarly, milk consumption has also increased more than 1.5 times during the same period.

The dairy sector however is not without its issues. Milk production is concentrated in ten states in India. The success of artificial insemination is around 35%. In addition, inter-calving interval and age at first lactation period are longer. Since the major players in the dairy industry are cooperatives, their financial health is important. However more than 50% of the cooperatives are under cumulative losses.

Availability of balanced rations and quality feed has been a constraint for many farmers. Indigenous cows yield about one-third of the yield of cross/exotic breed cows. This would require artificial insemination program to spread widely and availability of sex-sorted semen. In addition financing and insurance of cattle is a major issue as most financial institutions are not comfortable with it due to lack of a unique identity and methods to monitor the asset (cattle) that is financed.

In February 2020, the Government of India announced the scheme of promoting 10,000 farmer producer organizations by 2024. Both in agricultural sector in general and dairy sector, small and marginal farmers dominate. Therefore the major purpose of FPOs was to help restructure the value chain, reduce risk due to the risk pooling of the FPO members, and reduce transaction costs due to the availability of a business entity of FPO that represents several individual members who would otherwise have to conduct individual transactions. Provision of appropriate inputs and services through communication and economic sourcing and taking care of the economic welfare of the members are other benefits of the formation of an FPO.

Dvara E-Dairy Solutions Pvt. Ltd. is an agri-fintech firm that aims to empower especially small and medium dairy farmers with digital financial and cattle management solutions using an application. One part of the solution the organization provides is an electronic tag for animals called Surabhi e-Tag. This tag helps in cattle identification based on the muzzle identity with a great deal of accuracy. In addition, it has a mobile application named Surabhi Score. This score is a reflection of the animal's nutritional and reproduction status. This score is arrived at using a specified image of the cow and details obtained from the farmer. This application therefore acts as a diagnostic tool in assessing the health status and therefore helps in monitoring feed and other parameters accordingly. Thus Dvara E-Dairy has developed an IoT system that helps conserve resource use and increase animals' welfare. It uses real-time sensor data, machine learning technologies and cloud-based services which facilitate the dairy farmer. Sensors track the animal's fertility and health using its eating behavior, heat expression, and rumination 24/7. Using real-time data, analysis and information are conveyed through Internet to farmers' devices. Thus a combination of identity and health monitoring helps both farmer's decision-making and service provisioning across the dairy value chain.

Dvara E-Dairy has an opportunity to expand its business faster if it targets the FPOs as they represent a B2B business opportunity with members flowing suit. FPOs represent networks that with the help of digital technology can help small and marginal farmers to join the larger network that agritech or in this case dairy fintech firm like Dvara E-Dairy. Technology adoption is a function of several aspects. The FPO member's perception of the new technology for increasing the production and productivity of their milch animals depends on the perceptions of usefulness, ease of use, result demonstrability, visibility, and trialability. In each of these characteristics while an application like the Dvara E-Dairy needs to be successful, it requires some time for impacting on the perception of result demonstrability. In addition, there is a need to work on post-adoption attitudes too as image enhancement is one of the aspects. Dvara E-Dairy as a company was incorporated in 2019, has been in existence for only 2 years, and has been concentrating on the south and west of India for its operations. Given its limited time and geography, it would require efforts to create a database of adopters to convince new FPOs and its members to adopt technology.

The interviews with FPOs revealed that farmers have a choice with the use of technology. FPOs coordinate technology demonstration to farmers and negotiate a bulk discounted service charge for the agritech services; the choice however of using or not using the technology remains with the individual small dairy farmer. Another aspect of technology use is its characteristics and the nature of diffusion that depends on four aspects, namely, innovation, communication channels, time, and social system. In terms of innovation, its diffusion depends on relative advantage, compatibility, complexity, divisibility, and communicability. Relative advantage relates to the way dairy farmers can obtain benefits in feed and health management in the current form compared to the agritech-infused process. Compatibility refers to the use of the technology without affecting the current dairy farm management practices. Complexity refers to the difficulty in the use of technology. Divisibility refers to the bouquet of services which can be utilized individually compared to the entire bouquet. Communicability is the way that the benefits of the technology can be easily demonstrated.

Dvara E-Dairy introduced a service called Surabhi Mitra in response to the issue of technological complexity. These Surabhi Mitras are local part-time partners employed to ensure data collection on feed and health management to ensure appropriate recommendation for better adoption. This service however comes at a cost. One of the issues in this service is therefore the relative advantage when compared to the existing management practices followed by the dairy farmers who are the members of dairy FPOs. In addition, the effects of technology-based digital monitoring of the feed and health have its effects on a sustained basis only when it is continued. This continuance requires cost. The sustained monitoring also means that communicability is an issue for convincing FPOs and farmer FPO members to adopt technology. The use of technology is influenced by performance expectancy, effort expectancy, social influence, and enabling conditions. In this case, all these aspects become important.

The major issues in technology adoption also relate to FPOs. Women are still a minority in many FPOs resulting in their voice not being heard. Most members in a FPO would expect benefits like demand articulation, input service provision, and capacity building through training especially for use of technology and facilitating financing. Many FPOs are still in a nascent stage on all these counts. Moreover some of the issues plaguing FPOs include lack of adequate talent in managing the organization, operational issues, and weak governance. Data also shows that FPOs have expanded with government schemes and subsidies and come down when these are not available. Therefore, this observation clearly shows that FPOs are not keen organizationally to help its members. Another major issue with FPOs is the low paid-up capital that results in a substantial reduction in business activities, member interest, and patronage. While the top FPOs are dairies, these were converted from cooperatives to dairies, among the successful ones and not ones that have been successful in the very few years of existence. In the absence of substantial equity and fixed assets, most financial institutions are not keen on FPOs for lending.

7   Conclusions and Policy Implications

This study was conducted using a case research method to understand if agritech adoption and agritech companies can create the synergy in developing FPOs. This research used a dairy fintech company to understand the implications. The development and use of digital technologies to capture data for use in dairying can potentially help in increasing the efficiency and effectiveness of the dairy business. FPOs are organizational entities that can help in aggregating the buying and selling power of the small and marginal farmers in order to reap the benefits of the dairy fintech-related technologies. However, several limitations of these entities need to be addressed to achieve synergy with the use of dairy fintech technologies. These therefore result in policy implications.

Technology should not be looked at in isolation for bettering the farmer's condition. In the case of dairying even with the formation of a FPO, the share capital is very low; financial institutions are not interested in lending. This would require efforts in different ways to increase share capital. In the case of the majority of the FPOs that are small and marginal farmer collective, it requires additional funding for equity infusion through the government or other entities. In addition, this can be specifically encouraged if women manage the FPO.

Asset building is also another area that would help in access to finance. Sustainability and continuous use of technology depend on demonstrated benefits that are significant compared to the costs involved. This would mean that the top management of the FPO should proactively communicate and use their innovative members in highlighting the benefits of technology use. Government' efforts at increasing the availability of balanced ration for cattle development can also help.

The use of muzzle identity can be mandated across all the insurance companies. Cattle insurance can be propagated in a major way to attract more players into use of technology. Thus, it cannot only enthuse individual dairy farmers but also dairy producer companies to avail of insurance in a major way. The entry of insurance also would help monitor the health status of the animal in an organized way resulting in bettering of the sector. So supporting technology can also help in the adoption of these technologies.

Group performance follows through stages of forming, storming, norming, and performing. In the case of new FPOs, a key requirement therefore would be help with training for the members in developing a cohesive FPO with good coordination among its members. Government can also create awareness about the availability of new technologies for better tracking of feed and health management of milch animals among small farmers by initially subsidizing their minimal investment in technology use. Moreover, it can act on social influence as a factor to stimulate greater technology use. Demonstration farm, where the use of technology in the long run has resulted in greater benefits, is another way to help the spread of technology. The availability of technology in local languages can be encouraged especially with the perceived complexity due to the language barriers and low literacy among women who tend to cattle.

Acknowledgments The authors would like to thank the R&A Division, Ministry of Corporate Affairs, and the Government of India for funding this project under the research component of Corporate Data Management Plan Scheme (CDMPS).

References

Alagh, Y. K. (2019). Companies of farmers. In K. Amar & J. R. Nayak (Eds.), *Transition strategies for sustainable community systems: Design and systems perspectives*. Springer.

Amato, A., Amato, F., Angrisani, L., Barolli, L., Bonavolontà, F., Neglia, G., & Tamburis, O. (2021). *Artificial intelligence-based early prediction techniques in agri-tech domain*. In International conference on intelligent networking and collaborative systems (pp. 42–48). Springer.

Attuario, D., Bueno, V., Galasso, V., & Corghi, A. (2019). *Artificial intelligence application for the optimization of the supply chain: A real case experience for a leading global agritech company*. In 26th World Road CongressWorld Road Association (PIARC).

David, A. (2020). Consumer purchasing process of organic food product: An empirical analysis. *Journal of Management System-Quality Access to Success (QAS), 21*(177), 128–132.

David, A., Kumar, C. G., & Paul, P. V. (2022). Blockchain technology in the food supply chain: Empirical analysis. *International Journal of Information Systems and Supply Chain Management (IJISSCM), 15*(3), 1–12.

Deka, N., Goswami, K., Thakur, A. S., & Bhadoria, P. B. S. (2020). Are farmer producer companies ready to behave as business entities? Insights from the vegetable-based farmer companies in West Bengal, India. *International Journal of Agricultural Sustainability, 18*(6), 521–536.

Dolfsma, W., Isakhanyan, G., & Wolfert, S. (2021). Information exchange in supply chains: The case of agritech. *Journal of Economic Issues, 55*(2), 389–396.

Eisenhardt, K. M. (2021). What is the Eisenhardt method, really? *Strategic Organization, 19*(1), 147–160.

Finger, R., Huber, R., Wang, Y., Späti, K., & Ehlers, M. H. (2020). *How digital innovations can lead to more sustainable agricultural systems*. In 3rd INFER symposium on agri-tech economics for sustainable futures.

Ganeshkumar, C., David, A., & Jebasingh, D. R. (2022). Digital transformation: Artificial intelligence based product benefits and problems of agritech industry. In *Agri-Food 4.0*. Emerald Publishing Limited.

Ganeshkumar, C., Jena, S. K., Sivakumar, A., & Nambirajan, T. (2021). Artificial intelligence in agricultural value chain: Review and future directions. *Journal of Agribusiness in Developing and Emerging Economies*.

Ganeshkumar, C., Prabhu, M., & Abdullah, N. N. (2019). Business analytics and supply chain performance: Partial least squares-structural equation modeling (PLS-SEM) approach. *International Journal of Management and Business Research, 9*(1), 91–96.

Ganeshkumar, C., Prabhu, M., Reddy, P. S., & David, A. (2020). Value chain analysis of Indian edible mushrooms. *International Journal of Technology, 11*(3), 599–607.

Giaffreda, R. (2019). A dive into the AgriTech world: Technologies and adoption incentives. *IEEE Internet of Things Magazine, 2*(4), 44–45.

Gokul, V. U., Balaji, P., & Sivakumar, S. D. (2019). Role of actors in fFarmer Producer Organization (FPO) based millet value chain. *Madras Agricultural Journal, 106*(Special), 288–291.

Govil, R., Neti, A., & Rao, M. (2020). *Farmers producer companies: Past, present and future, report*. Azim Premji University.

Jose, E., & Meena, H. R. (2019). Profile of farmer producer company (dairy based) members in Kerala. *Indian Journal of Extension Education, 55*(2), 47–51.

Kakati, S., & Roy, A. (2019). Financial sustainability: A study on the current status of farmer producer companies in Northeast India. *IUP Journal of Management Research, 18*(2).

Kanitkar, A. (2016). The logic of farmer enterprises, Occasional Publication 17, Institute of Rural Management Anand (IRMA).

Kathiravan, N., Senthilkumar, T., & Senthilkumar, G. (2020). Farmers perception on the communication behaviour and usefulness of farmer producer organizations in Namakkal District of Tamil Nadu. *Asian Journal of Agricultural Extension, Economics & Sociology,* 9–13.

Kendall, H., Clark, B., Li, W., Jin, S., Jones, G., Chen, J., & Frewer, L. (2021). Precision agriculture technology adoption: A qualitative study of small-scale commercial "family farms" located in the North China plain. *Precision Agriculture,* 1–33.

Kenny, U., Regan, Á., Hearne, D., & O'Meara, C. (2021). Empathising, defining and ideating with the farming community to develop a geotagged photo app for smart devices: A design thinking approach. *Agricultural Systems, 194,* 103248.

Knierim, A., Borges, F., Kernecker, M., Kraus, T., & Wurbs, A. (2018). *What drives adoption of smart farming technologies? Evidence from a cross-country study.* In Proceedings of the European International Farm Systems Association Symposium (pp. 1–5).

Krishna, D. K. (2018). Farmer producer organizations: Implications for agricultural extension. *Agriculture Extension Journal.*

Kumar, P., Manaswi, B. H., Prakash, P., Anbukkani, P., Kar, A., Jha, G. K., & Lenin, V. (2019a). Impact of farmer producer organization on organic chilli production in Telangana, India. *Indian Journal of Traditional Knowledge (IJTK), 19*(1), 33–43.

Kumar, S., Thombare, P., & Kale, P. (2019b). Farmer Produce Company (FPC): A journey of farmers from food grower to food seller. *Agriculture & Food: e-Newsletter, 1*(9), 42–46.

Mahajan, V. (2014). Farmer producer companies: Need for capital and capability to capture the value added. In S. Dutta (Ed.), *State of India's livelihoods report 2014, access Livelihood services.* Oxford University Press.

Merriam, S. B., & Grenier, R. S. (Eds.). (2019). *Qualitative research in practice: Examples for discussion and analysis.* Wiley.

Pachayappan, M., Ganeshkumar, C., & Sugundan, N. (2020). Technological implication and its impact in agricultural sector: An IoT based collaboration framework. *Procedia Computer Science, 171,* 1166–1173.

Padmaja, S. S., Ojha, J. K., Shok, A., & Nikam, V. R. (2019). *Farmer producer companies in India: trends, patterns, performance and way forward.* National Institute of Agriculture Economics and Policy Research. Paper presented at Regional Conference on "Models for Agricultural Development:The experiences on Farmer Producer Companies (FPC)" organized by Department of Agricultural Economics and Indian Society of Agricultural Economics (ISAE) on 25th and 26th March 2019 at College of Forestry, Kerala Agricultural University Main campus.

Panpatte, S., & Ganeshkumar, C. (2021). *Artificial intelligence in agriculture sector: Case study of blue river technology.* In Proceedings of the second international conference on information management and machine intelligence (pp. 147–153). Springer.

Payne-Gifford, S., Johnson, K., Mauchline, A., Gadanakis, Y., Girling, L., & Mortimer, S. (2020). Exploring attitudes to technology adoption for cross compliance in Greek and Lithuanian farmers (No. 2308-2020-1586).

Prasad, C. S. (2017). *Framing futures.* In National Conference on Farmer Producer Organisations, Institute of Rural Management Anand (IRMA).

Raju, K. V., Kumar, R., Vikraman, S., Shyam, M., Rupavatharam, S., Kumara Charyulu, D., & Wani, S. P. (2017). *Farmer producer organization in Andhra Pradesh: A scoping study.*Rythu Kosam Project. Research Report IDC-16. ICRISAT.

Ramappa, K. B., & Yashashwini, M. A. (2018). Evolution of farmer producer organizations: Challenges and opportunities. *Research Journal of Agricultural Sciences, 9*(4), 709–715.

Serrano, P. V. H., Altenburg, L., & Kumar, P. (2020, December). *An exploratory analysis on agritech policies, innovations and funding for climate change mitigation.* In 2020 IEEE International conference on big data (big data) (pp. 2365–2370). IEEE.

Shah, T. (2016). Farmer producer companies: Fermenting new wine for new bottles. *Economic and Political Weekly, 51*(8).

Siddhartha, T., Nambirajan, T., & Ganeshkumar, C. (2019). Production and retailing of self-help group products. *Global Business and Economics Review, 21*(6), 814–835.

Siddhartha, T., Nambirajan, T., & Ganeshkumar, C. (2021). Self-help group (SHG) production methods: Insights from the union territory of Puducherry community. *Journal of Enterprising Communities: People and Places in the Global Economy.*

Simelton, E., & McCampbell, M. (2021). Do digital climate services for farmers encourage resilient farming practices? Pinpointing gaps through the responsible research and innovation framework. *Agriculture, 11*(10), 953.

Sowmya, V., & Raju, K. V. (2017). *Farmer producer organization profiles: Part-2. Rythu Kosam project.* Research Report IDC-16. Patancheru 502 324. International Crops Research Institute for the Semi-Arid Tropics.

Spanaki, K., Karafili, E., Sivarajah, U., Despoudi, S., & Irani, Z. (2021). Artificial intelligence and food security: Swarm intelligence of AgriTech drones for smart AgriFood operations. *Production Planning & Control,* 1–19.

Stake, R. E. (1988). Case study methods in educational research: Seeking sweet water. *Complementary Methods for Research in Education, 2,* 401–422.

Suriyapriya, E., Kavaskar, M., & Govind, S. (2019). Perception of the members of farmer producer organization on mobile agro-advisory service. *Journal of Global Communication, 12*(1), 1–5.

Trebbin, A. (2016). Producer companies and modern retail: Current state and future potentials of interaction. In N. Chandrasekhara Rao, R. Radhakrishna, Ram Kumar Mishra, & Venkata Reddy Kata (Eds.), *Organised retailing and agri-business: india studies in business and economics,* Springer.

Victer Paul, P., Ganeshkumar, C., Dhavachelvan, P., & Baskaran, R. (2020). A novel ODV crossover operator-based genetic algorithms for traveling salesman problem. *Soft Computing, 24*(17), 12855–12885.

Paul, V., Ganeshkumar, C., & Jayakumar, L. (2019). Performance evaluation of population seeding techniques of permutation-coded GA traveling salesman problems based assessment: Performance evaluation of population seeding techniques of permutation-coded GA. *International Journal of Applied Metaheuristic Computing (IJAMC), 10*(2), 55–92.

Verma, S., Singh, R., & Sidhu, M. S. (2017). A case study of selected farmer producer organization for promoting processed food in Punjab. *Indian Journal of Agricultural Marketing, 31*(1), 15–23.

Verma, S., Sonkar, V. K., Kumar, A., & Roy, D. (2019). Are farmer producer organizations a boon to farmers? The evidence from Bihar, India. *Agricultural Economics Research Review, 32*(conf), 123–137.

Vijayakumar, A. N. (2020). Coconut farmer producer organisation in socio-economic transformation of rural economy in Karnataka. *International Journal of Agricultural Resources, Governance and Ecology, 16*(3–4), 301–319.

Vishnu, S., & Gupta, J. (2017). Evolving localized innovation system: The case of Imasree farmer producer company. *Rural Extension and Innovation Systems Journal, 13*(2), 96–105.

Yin, R. K. (2018). *Case study research and applications: Design and methods.* SAGE Publications Inc.

Index

Printed in the United States
by Baker & Taylor Publisher Services